Mixed Methods Research Design for the Built Environment

The application of mixed methods research design in the built environment discipline by students and academics has continued to grow exponentially. However, with no dedicated mixed methods research design textbook in this domain, students have struggled to conduct research projects involving a mixed methods research design.

Mixed Methods Research Design for the Built Environment provides a useful research methodology resource for students, academics, and researchers across various disciplines in the built environment such as construction management and project management, property and real estate management, quantity surveying and commercial management, building surveying, building services engineering, civil and geodetic engineering, and other built environment disciplines. This book can also be useful for students and academics outside the built environment knowledge domain.

This textbook offers practical and step-by-step guidance on how to apply mixed methods research design, including an elucidation of the various philosophical and methodological underpinnings upon which the choice of a particular variant of the mixed methods research design is predicated. It provides practical case examples and guidance on the processes involved to design and undertake mixed methods research, the advantages and disadvantages of using mixed methods research, and how multiple sources of qualitative and quantitative data can be combined and applied to carry out research projects.

Julius Akotia is a senior lecturer in Construction Project Management in the School of Architecture, Computing and Engineering (ACE), University of East London (UEL). He has a strong interest in research and has participated in major research projects, such as the London Olympic project with the CIOB team of researchers. His research interests lie in the area of sustainability.

Bankole Osita Awuzie is an associate professor in Construction Management at the School of Construction Economics and Management at the University of Witwatersrand, Johannesburg, South Africa. Dr Awuzie's research interests are situated along a multi-, inter-, and transdisciplinary (MIT) plane with a focus on the facilitation of sustainable, smart, and circular built environments in developing country contexts.

Charles Egbu is a professor and the vice chancellor, Leeds Trinity University, England, United Kingdom. His research interests are in construction project management, mental health issues in construction, sustainable development, innovations and knowledge management in complex environments, and research methodology and methods.

Mixed Methods Research Design for the Built Environment

Edited by
Julius Akotia, Bankole Osita Awuzie, and Charles Egbu

LONDON AND NEW YORK

Designed cover image: "City Urban Development and Nature"
by Geralt on Pixabay.com

First published 2024
by Routledge
4 Park Square, Milton Park, Abingdon, Oxon, OX14 4RN

and by Routledge
605 Third Avenue, New York, NY 10158

Routledge is an imprint of the Taylor & Francis Group, an informa business

ISBN: 9781032068268 (hbk)
ISBN: 9781032065595 (pbk)
ISBN: 9781003204046 (ebk)

DOI: 10.1201/9781003204046

Typeset in Times New Roman
by codeMantra

Contents

Figures

Tables

Abstract

The increasing use of mixed methods research design within the built environment discipline has been observed in recent times. Students at undergraduate and postgraduate levels, as well as their academic supervisors, are increasingly recognising the importance of applying mixed methods research design as the best-fit design, most suitable for conducting research into complex human activities in built environment disciplines. They have continued to see the importance of combining multiple research methods as a viable alternative research method in addressing complex and contemporary research questions. However, there is currently no textbook that can provide a detailed guidance on the application of various typologies of the mixed methods research designs within the built environment discipline using examples that span across various facets of the knowledge domain such as construction economics, disaster management, construction procurement, construction health and safety, architectural technology education, construction skills and competences, and sustainability. Hence, this textbook offers a practical and step-by-step guidance on how to apply mixed methods research design across the built environment disciplines, including an elucidation of the various philosophical and methodological underpinnings upon which the choice of a particular mixed methods research design typology is predicated. It provides practical case examples and guidance on the processes involved to design and undertake a mixed methods research, the advantages and disadvantages of using mixed methods research, and how multiple sources of qualitative and quantitative datasets can be combined and applied to carry out research projects in the built environment domain. The mixed methods research design typologies that have been covered include exploratory and explanatory sequential, concurrent, or convergent and embedded or nested mixed methods designs. This textbook serves as a useful research methods textbook for undergraduate and postgraduate students in construction management and project management, quantity surveying, building surveying, building services engineering, civil and geodetic engineering and surveying, construction law, property and real estate management, and other built environment discipline.

Editor Biographies

Julius Akotia is a senior lecturer in Construction Project Management in the School of Architecture, Computing and Engineering (ACE), University of East London (UEL). He is also the link coordinator for the department's international academic partnership programmes for construction management and civil engineering BSc/BEng and MSC programmes with Metropolitan College in Greece and Ain Shams University in Egypt. He has taught and managed construction management courses at Kingston University, London and the University of Salford. He is a co-editor of the *Secondary Research Methods in the Built Environment* textbook. He has supervised several undergraduate, postgraduate, and doctorial students and played external examiners' role for national and international institutions. Prior to joining academia, he has practised as Quantity Surveyor in the UK construction industry on several construction projects. He is a chartered member of Chartered Institute of Building (MCIOB) and a senior fellow of the Higher Education Academy (SFHEA). Julius holds a PhD in Construction Project Management from the University of Salford, Manchester. He has a strong interest in research and has participated in major research projects, such as the London Olympic project with the CIOB team of researchers. His research interests lie in the area of sustainability. He has published several research papers in local and international conferences and academic journals.

Bankole Osita Awuzie is an associate professor in Construction Management at the School of Construction Economics and Management at the University of Witwatersrand, Johannesburg, South Africa. Also, he is an adjunct lecturer in the School of Environmental Sciences at K.O. Mbadiwe University, Ogboko, Imo State, Nigeria. He holds a PhD in Built Environment with an emphasis on Strategic Construction Procurement from the University of Salford, Salford, an MSc in Construction Project Management from Robert Gordon University, Aberdeen, Scotland, United Kingdom, and a BSc (Hons) degree in Estate Management from Imo State University, Owerri, Nigeria. Dr Awuzie' s research interests are situated along a multi-, inter-, and transdisciplinary (MIT) plane with a focus on the facilitation of sustainable, smart, and circular built environments in developing country contexts. Besides participating in various research projects as principal and co-investigator respectively, Bankole has authored and co-authored more than 100 peer-reviewed publications comprising journal articles in leading built environment and MIT-oriented journals, conference papers, book chapters, and two books. To date, Bankole has supervised and co-supervised four doctoral and 11 masters by research degree candidates to successful completion. Dr. Awuzie is a Y2-rated researcher of the South African National Research Foundation (NRF). He serves as an associate editor with the *Built Environment Project and Asset Management* (BEPAM) and Frontiers in *Sustainable Organizations* Journal as well as a member of the Editorial Board of *Sustainable Development.*

Charles Egbu is a professor and the vice chancellor, Leeds Trinity University, England, United Kingdom. He was Pro-Vice Chancellor (Education & Experience) at the University of East London, England, United Kingdom. Before then, he was the dean of the School of the Built Environment and Architecture, London South Bank University, United Kingdom, where he held the chair in Project Management and Strategic Management in Construction. He has over 25 years of experience in the UK Higher Education Sector. His first degree was in Quantity Surveying [First Class Honours, United Kingdom]. His doctorate was obtained from the University of Salford, United Kingdom, in the area of Construction Project Management. He was President of the Chartered Institute of Building (June 2019–June 2020). He is an author of over 12 books on Project Management, Construction Management, Knowledge Management, and Sustainable Development. He has contributed to over 350 publications in various international journals and conferences and has supervised over 30 PhD students and examined over 100 PhD candidates worldwide. He has managed several large multi-disciplinary research projects (received over £25m funding) as Principal Investigator and led a number of successful Research Centres. He was formerly a member of the Peer Review College of the UK Economic and Social Sciences Research Council (ESRC) and a member of the Peer Review College of the UK Engineering and Physical Sciences Research Council (EPSRC). His research interests are in construction project management, mental health issues in construction, sustainable development, innovations and knowledge management in complex environments, and research methodology and methods.

Contributors

Henry Abanda is a reader in the School of the Built Environment, Oxford Brookes University. His research interests are in the area of Semantic Web, BIM, and Big Data. He has worked on research projects funded by the Engineering & Physical Sciences Research Council, the International Labour Organisation, and the Intergovernmental Panel on Climate Change.

Samuel Adekunle obtained his PhD in construction management from the University of Johannesburg. He has several publications in accredited peer-reviewed journals, book chapters, and conferences, which have earned him recognition. Samuel has been involved in several research projects in different capacities. He has served as a panel member on several platforms and seminars discussing the construction industry improvement strategies.

Douglas Aghimien is a vibrant researcher with a keen interest in construction digitalisation, construction performance, value management, and sustainable construction. He has a PhD in engineering management from the University of Johannesburg, South Africa and has published over 100 peer reviewed articles, books, and book chapters. Also, he is a deputy editor of the *Journal of Construction Project Management and Innovation*.

Clinton Aigbavboa is a professor of Sustainable Human Development in the Department of Construction Management and Quantity Surveying, University of Johannesburg, South Africa. He is an active postgraduate degree supervisor (master's and PhD), with over 100+ master's and 45 doctoral students supervised to completion.

Julius Akotia is a senior lecturer in Construction Project Management in the School of Architecture, Computing and Engineering (ACE), University of East London (UEL). He has taught construction management courses at Kingston University, London, and the University of Salford. He has also, over the years, carried out module leadership and supervision roles for several construction management modules. He is a co-editor of the *Secondary Research Methods in the Built Environment* textbook. He has supervised several undergraduate, postgraduate, and doctorial students and played external examiners' role for national and international institutions. Prior to joining academia, he has practised as a quantity surveyor in the UK construction industry on several construction projects. He is a chartered member of Chartered Institute of Building (MCIOB) and a senior fellow of the Higher Education Academy (SFHEA).

Bankole Osita Awuzie is an associate professor in Construction Management at the School of Construction Economics and Management at the University of Witwatersrand, Johannesburg, South Africa. His current research interests focus on the facilitation of sustainable, smart, and circular built environments. He has authored and co-authored several peer-reviewed publications, including two books focusing on this research area. Also, he has supervised several postgraduate studies at masters' and doctoral degree levels to successful completion.

Mohd Azrai Azman is a senior lecturer in the Faculty of Architecture, Planning and Surveying at the Universiti Teknologi MARA, Sarawak Branch, Kota Samarahan, Sarawak, Malaysia. His current research interests include productivity issues and economic analysis in the construction sector. He has published articles in the *International Journal of Project Management*; *Construction Management and Economics*; *Engineering, Construction and Architectural Management*; and the *Journal of Management in Engineering*.

Souad Sassi Boudemagh is a professor of Architecture in the Project Management Department of the University of Constantine 3. Souad's research interests are in educational evaluation, higher education, and educational management. Their most recent publication is "Identification of Factors Causing Delays in Construction Projects in Algeria".

Sandra Carrasco is a CIFAL postdoctoral research associate at the University of Newcastle, Australia. She was awarded the McKenzie Postdoctoral Fellowship at the University of Melbourne. Sandra holds master's and PhD degrees in Environmental Management from Kyoto University, sponsored by the Japan Government's MEXT scholarship. Her research interests include post-disaster reconstruction and community resilience, disaster risk reduction, incremental housing, governance and city planning, and refugee studies.

Shamy Yi Min Chin is a contract executive with five years of working experience, currently working at a contractor firm based in Penang. She specialises in cost estimation, contract administration, and final account. She strives to ensure that every project is completed within the client's budget while ensuring the company's profitability.

George Denny-Smith is a Scientia PhD researcher at UNSW Sydney who grew up and still lives on the traditional lands of the Bidjigal and Gadigal peoples of the Eora Nation in Warrang, Sydney. George works with First Nations organisations to examine the social value of Indigenous procurement policies in construction.

Andrew Ebekozien is a lecturer in the Department of Quantity Surveying, Auchi Polytechnic, Nigeria. He obtained his PhD in Cost Management from Universiti Sains Malaysia, Malaysia. He is the author/co-author of many peer-reviewed journal articles.

Temitope Egbelakin is a professor at the University of Newcastle, Australia. Dr. Egbelakin is an international scholar and leader with significant experience in teaching, research, and industry practice in multiple local and international contexts. Her interests and expertise include Disaster Resilience, Smart and Resilience Cities, Informatics, and Maintenance and Adaptive Reuse of Buildings.

Charles Egbu is a professor and the vice chancellor, Leeds Trinity University, England, United Kingdom. He was Pro-Vice Chancellor (Education & Experience) at the University of East London, England, United Kingdom. Before then, he was the dean of the School of the Built Environment and Architecture, London South Bank University, United Kingdom, where he held the chair in Project Management and Strategic Management in Construction. He has over 25 years of experience in the UK Higher Education Sector. He was President of the Chartered Institute of Building (June 2019–June 2020). He is an author of over 12 books on Project Management, Construction Management, Knowledge Management, and Sustainable Development. He has contributed over 350 publications in various international journals and conferences and has supervised over 30 PhD students and examined over 100 PhD candidates worldwide. He has managed several large multi-disciplinary research projects (received over £25m funding) as Principal Investigator and led a number of successful Research Centres. He was formerly a member of the Peer Review College of the UK Economic and Social Sciences Research Council (ESRC) and a member of the Peer Review College of the UK Engineering and Physical Sciences Research Council (EPSRC). His research interests are in construction project management, mental health issues in construction, sustainable development, innovations and knowledge management in complex environments, and research methodology and methods.

Obuks Ejohwomu is an associate professor (senior lecturer) in Project Management at the University of Manchester. He is an award-winning research-active academic with a world-leading profile for delivering impact-driven research in construction engineering and management, particularly in digitalisation, productivity, health and safety, and the future sustainable built environment. In this capacity, he has co-authored over 82 publications.

Duga Ewuga is a quantity surveyor and currently lectures at the Liverpool John Moores University, United Kingdom. He undertook his PhD on developing a framework for sustainable procurement practice for Irish construction firms. He was a Fiosraigh doctoral scholar at the Technological University Dublin.

Gabriela Fernandes is an assistant professor at the Faculty of Sciences and Technology, University of Coimbra. Her research interests are in Organisational Project Management and Innovation Management, particularly in University-Industry R&D Collaborations context. She spent ten years in the coordination and management of projects in different industries and is author of more than 100 publications in journals and conferences.

Cheng Siew Goh is a professional surveyor and academic. Her subject knowledge lies in sustainability, green buildings, life cycle management, construction project management, construction innovations, BIM, mixed reality, 3D scanning, and robotics. She has undertaken individual and joint research projects to produce high-quality research outputs, and her works have been published in peer-reviewed journals, conference proceedings, and books. Her expertise is highly recognised and she has been invited to serve as Reviewer for several peer-reviewed journals and international conferences.

Nishani Harinarain is Associate Professor at the University of KwaZulu-Natal and has been in academia for 15 years. She had taught and supervised numerous students. Prof Harinarain has received numerous awards such as the Distinguished Teachers Award and the Women in Engineering and the Built Environment.

Abid Hasan is a lecturer in the School of Architecture and Built Environment at Deakin University, Australia. In this role, he supervises both undergraduate and postgraduate research in the built environment discipline. His research interest includes human-technology interaction, occupational health and safety, project success factors, and productivity in construction projects.

Carol K. H. Hon is a senior lecturer and research fellow (DECRA) at the School of Architecture and Built Environment, Faculty of Engineering, Queensland University of Technology, Australia. She is passionate about reducing construction accidents and enhancing safety performance, publishing a research-based textbook *Safety of Repair, Maintenance, Minor Alteration, and Addition (RMAA) Works*. Her research, funded by the Centre of Workplace Health and Safety, contributes to developing an innovative training program to reduce electrical risks for young construction workers. She uses Bayesian and agent-based modelling techniques in her DECRA project, delivering evidence to better manage psychosocial risks and reduce suicide rates in the construction industry.

Alan Hore has over 30 years of experience as a chartered surveyor and academic manager. He is the assistant head of the School (Head of Quantity Surveying Programme) of Surveying and Construction Management at the Technological University Dublin, Republic of Ireland. Alan is also the founder of the Construction IT Alliance (CITA) in Ireland.

Nigel Isaacs is a senior lecturer in the School of Architecture, Victoria University of Wellington. His research interests include post-occupancy evaluation, the development of the building technologies used in New Zealand, and the use of energy and water in residential and non-residential buildings.

Iniobong Beauty John is a construction economics and construction management specialist with an interest in sectorial and national competitiveness, industrial economics, circular economy, informality, sustainable development, digitalisation, and commercial management. She has mentored over 50 postgraduate and undergraduate students through direct project supervision. She is a senior research associate at the University of Johannesburg, a senior lecturer at the University of Lagos, and co-cluster manager for the Urban Design and sustainable infrastructure cluster at the Centre of Excellence in Sustainable Housing and Habicities of African Research University Alliance.

Abdallah Lalmi is a doctor in Project Management and a temporary teacher at the University of Salah Boubnider Constantine 3 (Algeria). He has published several articles in international and national journals and conferences. His research interests include lean management, agile, and waterfall methods.

Champika Liyanage is a professor in the School of Engineering, UCLan and is the co-director of UCLan's Centre for Sustainable Transitions. Champika is actively involved in a wide array of research relating to sustainability, facilities and infrastructure management, and capacity building in disaster resilience. She has published over 200 peer-reviewed publications (journal and conference papers) to date. Through sheer hard work and perseverance, she was able to access a range of opportunities and experiences, such as developing a strong international research portfolio. By securing funding, including from Horizon 2020 and EU ERASMUS+, Champika has been able to carve a research path, developing a range of internal, national, and international projects and networks.

Martin Loosemore is Professor of Construction Management in the Faculty of Design, Architecture and Building at UTS. Social procurement is one of his areas of interest, and he has founded a successful social enterprise that specialises in securing employment opportunities in the construction and engineering industries for disadvantaged people.

Robert A. Lukan holds a Bachelor of Science Honours in Quantity Surveying Degree from the University of KwaZulu-Natal. He is an enthusiastic, diligent person who is currently working in a Quantity Surveying Consulting Firm.

Patrick Manu is Professor of Innovative Construction and Project Management at the School of Architecture and Environment, University of the West of England. He is a research active academic with international reputation for construction safety and health research, which has underpinned exceptional contribution to knowledge transfer and external engagement in the construction industry, both in the United Kingdom and internationally. He has been involved as principal investigator (PI) and co-investigator in research projects (valued at over GBP £1.8 million) funded by several organisations. He led (as PI) an international consortium in an EPSRC-funded research to develop the first web-based application for assessing design for safety organisational capability, which won an innovation award from HS2 Ltd. He has over 120 publications.

Udochukwu Marcel-Okafor, PhD, is a principal lecturer in the Department of Architectural Technology at Federal Polytechnic Nekede, Nigeria. She has co-authored various empirical studies. Her research interests are in architectural technology and education, especially the development and implementation of effective curriculum for the execution of sustainable, safe, and inclusive human settlements.

Mark Mulville is a building surveyor and architectural technologist and formerly an academic leader at the University of Greenwich, United Kingdom. He is currently the head of the Surveying and Construction Management School at the Technological University Dublin, Republic of Ireland.

Bevan Naidoo holds a Bachelor of Science Honours in Quantity Surveying Degree from the University of KwaZulu-Natal. He is a hardworking and meticulous individual, who is currently working in a Quantity Surveying Consulting Firm.

Stanley Njuangang is Course Director for Construction Management and Facilities Management at Leeds Beckett University. He is also the leader of the undergraduate dissertation for surveying and construction-related courses. His research interests are in healthcare facilities management and infection control, maintenance management, public–private partnerships (PPP), performance measurement and management. In addition to publishing peer-reviewed articles, Stanley has presented his work in many international conferences.

Beth Noble is a PhD candidate in Building Science in the School of Architecture, Victoria University of Wellington. Her research focuses on accessible built environments, particularly relating to the sensory environment for autistic users with a focus on electric lighting, as a member of the autistic community herself.

Jemima Antwiwaa Ottou holds a PhD in Construction, an MSc in Engineering and Management, and a BSc in Building Technology. She has over 20 years of functional experience in practice and more than eight years as faculty. Her research interests include Total Quality Management and Procurement in Construction Project Delivery.

Leanne Piggott is Associate Professor and WIL Central Academic Director at UNSW. As the National Education Director at the Centre for Social Impact in the UNSW Business School, she is responsible for leading curriculum quality and innovation. Leanne's professional and research interests are in social impact and shared value.

Chris Pye is a lecturer in Building Surveying in the School of Engineering at the University of Central Lancashire. Chris lectures on the topics of low- and high-rise construction technology and building pathology combined with facilities management. Chris' research interests include the application of building information modelling to construction applications and the conservation of military structures and war memorials.

Saeed Rokooei is an assistant professor in the Department of Building Construction Science at Mississippi State University. His professional responsibilities include project management as well as architectural design practice in private and public construction and engineering firms. He has taught architecture and construction programmes since 2006. His research interests include simulation, engineering education, project management methodologies, and data analytics.

Enoch Sackey is a senior lecturer in Project Management in the Department of Management of Nottingham Business School. He is an industry expert in the procurement of construction works, quantity surveying, and contract management. His current research interest centres on technological innovation in construction and sustainability in procurement.

Riza Yosia Sunindijo is Associate Professor of Construction Management UNSW Built Environment. Previously he worked as a project engineer, project manager, and sustainability champion in multi-national construction and project management organisations. He is drawn to the dynamic nature of construction where various stakeholders collaborate to achieve common project objectives.

Megan Williams is a Wiradjuri professor of Indigenous Health in the School of Public Health, Faculty of Health at UTS. As Head of Girra Maa, the Indigenous Health Discipline in the School of Public Health, she researches Aboriginal and Torres Strait Islander health and wellbeing, and human rights/justice issues.

Preface

The idea and motivation for this book culminated from the editors' recognition of the absence of a mixed methods research design textbook, detailing built environment exemplars to guide students and their supervisors within the built environment discipline who are desirous of carrying out research using any of these designs. From the experience of the editors as senior academics and research supervisors over the years, as well as the views they have received from the interactions with other academics across various HE institutions in the United Kingdom and globally, the absence of a dedicated mixed methods research design textbook in the built environment which provides guidance for students and their supervisors has continued to pose a challenge. This book sets out to address such a challenge.

Students at undergraduate and postgraduate levels as well as their academic supervisors are increasingly recognising the importance of applying mixed methods research designs as the best fit and have continued to see the importance of combining multiple research methods as a viable alternative research method in addressing complex and contemporary research questions within the built environment domain. The contemporary built environment research projects are becoming complex and difficult than they were years ago due to the complex and evolving nature of the built environment. Accordingly, the application of mixed methods research design in studies within the built environment discipline has continued to grow exponentially.

This textbook has a simple and ambitious goal. It offers a practical and step-by-step guidance on how to apply mixed methods research across the built environment disciplines, including an elucidation of the various philosophical and methodological underpinnings upon which the choice of a particular variant of the mixed methods research design is predicated. In particular, it provides practical case examples and guidance to illustrate the processes involved and how multiple sources of qualitative and quantitative data can be combined and applied towards carrying out research projects in the built environment domain.

The core audience for this textbook is students and academics within the built environment disciplines, including project management, construction management, quantity surveying, building surveying, building services engineering, civil and geodetic engineering, property and real estate management, and other built environment discipline areas.

This book is designed and structured in a manner that makes it easy for readers to understand, follow, and apply the case examples. It is suitable for both users who are new to research – who want to use it to learn the act and rudiment of mixed methods research design while those who are more advanced in research may use it to refresh their knowledge on the use of mixed methods research design. It can also serve as a reference source for users.

Throughout this book, an effort is made to structure the chapters in ways that facilitate the understanding of readers. By following the mixed method research design processes drawn from many case examples, this book enables novice researchers to develop a good appreciation of the main typologies of mixed methods research design.

While the focus of this textbook is on the built environment, the methodological processes and case examples presented can be applied to the wider audience in the social science research community. Hence, it is envisaged and hoped that the use of this book will not be limited to students and academics within the built environment but also be useful in the wider academic and social science research environment.

Julius Akotia, Bankole Osita Awuzie and Charles Egbu

1 Introduction to Mixed Methods Research Designs in the Built Environment

Julius Akotia, Bankole Osita Awuzie, and Charles Egbu

Summary

Mixed methods research design has continued to gain prominence as a veritable alternative research design to the traditional single research methods among built environment researchers. Owing to the complex and multi-faceted nature of the issues being understudied through contemporary built environment research, the utility of multiple research methods in offering researchers in this discipline the opportunity to investigate such issues more extensively. The increasing use of mixed methods within the built environment field highlights its importance and recognition as a viable research design by many built environment researchers. This chapter provides a background into the evolution of mixed methods research design within the built environment context, as well as the principle and theory behind its application. It details the various mixed methods research design typologies available to built environment researchers. Further, the chapter articulates the advantages and disadvantages of the design. The chapter enables readers to develop a broader understanding of the mixed research methods design, the motivation for mixed methods research and the integration processes involved. Summarily, the chapter provides an outline of the subsequent chapters.

Mixed Methods Research Design

There was a deliberate use of multiple research methods "long before anyone had identified this as a particular type of research" (Maxwell, 2016: 16). In recent times, mixed methods research has assumed prominence and acceptance as a feasible and viable alternative to the traditional, single, qualitative or quantitative, research method variants (Hanson *et al.*, 2005). According to Creswell and Garrett (2008), the recent demand for a mixed methods approach stemmed from concerns about the inability of the traditional, individual qualitative and quantitative research paradigms to offer workable solutions to the increasingly complex and dynamic problems confronting society and the world at large. The emergence of mixed methods research provided an alternative to the mono-methods which had been reported as being non-responsive to the ever-increasingly complex and multi-faceted challenges confronting humanity. Mixed methods research is now regarded

DOI: 10.1201/9781003204046-1

as "the most promising research method" for social and behavioural science studies (Brown, 2021: 4) across many disciplines. Several factors have accounted for the evolution and acceptance of mixed methods as a research design (Creswell, 2009). Complexity and diversity of contemporary research problems appear to be the most cited factors influencing the perpetuation of mixed methods research. Prior to its adoption and acceptance as a research method, it was traditionally used mainly in the fields of anthropology and sociology (Johnson *et al.*, 2007). The growing number of mixed methods research applications across "disciplines signal the advancing acceptance of mixed methods research in practice" (Plano Clark, 2010: 430).

There are several definitions of mixed methods research. For example, Johnson *et al.* (2007: 123) defined mixed methods research as "the type of research in which a researcher or team of researchers combine elements of qualitative and quantitative research approaches (e.g. viewpoints, data collection, analysis, inference, techniques) for the broad purposes of breadth and depth of understanding and corroboration". The diversity inherent in its design and definition indicates that mixed methods research has become critical, and is synonymous with good practice in research. According to Greene (2008: 20), the mixed methods research design "offers deep and potentially inspirational and catalytic opportunities to meaningfully engage with the differences that matter in today's troubled world". Underlying these definitions is the recognition of the design's uniqueness and ability to offer multi-dimensional research solutions to humanistic and behavioural phenomena in a manner that a single form of research method is unable to do. A significant proposition of mixed methods research is the diversification of ideas it offers as a method, together with its potential to broaden the understanding of human experiences in developing policies and practices (Tashakkori & Teddlie, 2010). Advancing the potential benefit argument, Greene (2008) cited corroboration and complementarity as some of the major advantages that are associated directly with the mixed methods research design.

Contemporary mixed methods research is also seen as the "third force or methodological paradigm" by authors like Combs and Onwuegbuzie (2010), and Tashakkori and Teddlie (2010), because it draws its strength and validity from the traditional qualitative/quantitative research methods and integrates them in a manner that helps to answer unique research questions pertaining to the scientific and social world (Klingner & Boardman, 2011). A major advantage of the mixed methods research design lies in its strong ties with research questions (Creswell & Garrett, 2008). According to Bryman (2006) and Hanson *et al.* (2005), the decision to adopt a mixed methods research design must be based on a number of reasons, notably the purpose of the study, the research questions, and the type of data required for the study. Underlying such determination is the rationale behind the use of mixed methods in providing the best research design to answer inductive-based and deductive-based research questions together within a single study. It is believed that the effective application of mixing two research methods will yield better research outcomes than can be achieved through a single research method (Johnson *et al.*, 2007). For example, combining interviews with a questionnaire survey can help to delve further into participants' knowledge and experience, yielding powerful insights for the study (Johnson & Onwuegbuzie, 2004). Equally, the fundamental principle behind mixed methods enables the researcher to collect data from multiple sources to investigate hard and soft societal issues without compromising the scientific rigour of the findings (Masadeh, 2012). Saunders *et al.* (2009) and Onwuegbuzie and Johnson (2006) argued that by adopting qualitative and quantitative research methodologies and methods (e.g. interviews and surveys) within the same research framework, practical questions can be addressed simultaneously from different perspectives, leading to greater confidence in the findings and conclusions. In addition, adopting a mixed methods design will enable the researcher to mix and match design elements in a way that provides the best opportunity to address specific research questions within a single study.

However, the fundamental question is how mixed methods research can be designed to ensure that the weaknesses of one research methodology (e.g. qualitative) are well complemented by the strengths of another research methodology (e.g. quantitative). In view of this, Onwuegbuzie and Johnson (2006) cautioned researchers adopting the mixed methods research design to examine carefully the extent to which the weaknesses and strengths from both methodologies can be counter-balanced without compromising the rigour and validity of the findings. Given the distinctive differences between the elements of qualitative-inductive and quantitative-deductive-oriented methodologies, Newman and Hitchcock (2011) advised researchers to focus on the purpose of the research to drive the method rather than focusing just on the strengths/weaknesses parameters and philosophical assumptions of combining the two methodologies.

Table 1.1 shows the strengths and weaknesses of a mixed methods research approach.

Mixed methods research is generally aligned with the pragmatist philosophy (Onwuegbuzie & Johnson, 2006). The advantages of combining the elements of qualitative and quantitative research, despite the differences in their philosophical

Table 1.1 Strengths and weaknesses of mixed methods research design

Strengths	*Weaknesses*
Words, images, and descriptions can be used to supplement the meaning of figures and vice versa.	Can be more expensive to conduct.
Stronger evidence can be provided through convergence and corroboration of findings.	Mixing two or more research paradigms can be difficult and problematic.
Can provide a broader perspective to a range of research questions and issues.	Can be time consuming than the single method approach.
Can offer deeper insights and understanding than the single approach method.	Can be difficult to analyse and draw inferences to interpret findings.
Can offer a more complete knowledge necessary to inform theory and practice.	Can generate a large volume of information/data.
The strength of one method can counter or overcome the weaknesses of the other method.	May require skills as one researcher may not be skilled in both methods.

Adapted from Johnson and Onwuegbuzie (2004).

orientations, have been acknowledged in the literature (Grix, 2004). Numerous questions have also been raised about the fundamental issues relating to its philosophical orientations. Previous contributors have argued that the philosophical barriers between the two methods, together with their contrasting views, made their (quantitative and qualitative) elements incompatible to combine. Moreover, according to Onwuegbuzie and Johnson (2006: 59), the combination of the two viewpoints has also been considered to be tenuous because of their "competing dualisms: epistemological (e.g. objectivist vs subjectivist), ontological (e.g. single reality vs multiple realities), axiological (e.g. value free vs value-bound), methodological (e.g. deductive logic vs inductive logic), and rhetorical (e.g. formal vs informal writing style) beliefs" that they espouse. However, while these research methodologies seem to be espousing different philosophical ideologies, they tend to provide a research approach and philosophical dimension that seeks to bring together their perspectives into a workable solution (Johnson & Onwuegbuzie, 2004). It can be argued that both qualitative and quantitative research methodologies have common, acknowledged elements that transcend these differences and barriers (Bryman, 2006). In practical terms, there are apparent overlaps between them, to some extent, which plays down the "difference" argument perceived to be existing between them. For this reason, by de-emphasising their philosophical differences (Chen, 2006) and aiming solely at their potential benefits, the perceived differences between the two methodologies are relegated, to a large degree, to the background.

In the built environment, researchers have since recognised the importance of combining qualitative-textual and quantitative-numerical-oriented research methods for their research projects. As such, mixed methods research has gained popularity among researchers in the built environment in recent times (Day & Gunderson, 2018). Panas and Pantouvakis (2010: 79) suggested that the mixed methods research concept "seems to be gaining ground, especially given the industry's change towards intensifying the exploration of productivity's soft aspects as well as behavioural and managerial factors and cultural diversions of the project actors". According to Amaratunga *et al.* (2002) and Day and Gunderson (2018), mixed methods design has been regarded as an emerging area of research, which has several advantages, particularly within the built environment discipline. While the traditional elements of quantitative enquiry are based on deductive reasoning, statistical analysis, and hypothesis testing, the traditional elements of qualitative inquiry, on the other hand, are based on inductive reasoning and theory generation. Given that construction processes are fundamentally complex with diverse players and rapid technological changes, at the centre of the exploration of these processes and complexities are the crucial roles that deductive and inductive reasoning play in ensuring the successful exploration of these issues. In this regard, the application of the traditional, single, research method, such as a quantitative- or qualitative-oriented research methodology, no longer appears to be adequate and suitable in dealing with the complex issues (soft and hard) that are usually associated with built environment research. There are practical benefits to be realised from using a mixed methods research design if researchers position themselves to understand the rationale of combining the elements of qualitative and quantitative research methods (Maxwell, 2016) for research in the built-environment discipline.

Evidence to date suggests that the use of a single research method approach exclusively has proven to be inadequate in exploring situations where the issues are of a multi-faceted nature, such as those found in the built environment domain where the interaction among processes and participants in projects remains a key feature, often requiring a substantial number of procedures. In these cases, investigating such complex inter-relations and interactions requires gathering substantial evidence (Creswell & Garrett, 2008). Similarly, since built environment activities are not discrete events but processes with different phases, involving different types of activities predominating at different times and levels, it, therefore, stands to reason that some particular research methodologies and methods might be more useful for some activities than others. Apparently, the combination of the relative strengths from multiple research methods has the potential to offer a more comprehensive and desirable outcome (Mingers, 2001). The application of such a combined approach will allow for both deductive and inductive reasoning and better appreciation of a given situation "rather than a strictly positivistic or interpretivist slant to the data" (Harrison & Reilly, 2011: 22). In the built-environment domain which primarily involves multi-disciplinary teams with enormous research challenges, combining different data sets and strategies from multiple sources (quantitative and qualitative) will enhance the reliability and the practical significance of the findings.

Mixed Methods Research Design Typologies

Numerous typologies of mixed methods research design have been developed. However, the most commonly used typologies are exploratory design, explanatory design, concurrent or convergent design, and embedded or nested design (Harrison & Reilly, 2011; Klingner & Boardman, 2011). For each of these typologies, the research processes followed, from research design through to data collection, analysis, and presentation, differ.

For exploratory sequential mixed methods design, also known as a two-phase design, the collection and analysis of the qualitative data is undertaken first and is followed by the collection and analysis of the quantitative data. The findings of the qualitative data are used to provide insight to inform subsequent quantitative data collection and analysis (Creswell & Hirose, 2019; Klingner & Boardman, 2011). The core principle of this design is to use the quantitative results to provide insight into the qualitative findings (Creswell, 2009). In this design, the priority or weighting is typically given to the qualitative data. The quantitative data is largely used to complement the qualitative data (Hanson *et al.*, 2005). This type of mixed methods design is useful where the researcher wants to explore "relationships when study variables are not known, refining and testing an emerging theory, developing new psychological test/assessment instruments based on an initial qualitative analysis, and generalizing qualitative findings to a specific population" (Hanson *et al.*, 2005: 229).

Conversely, in explanatory sequential mixed methods design (also two-phase design), the quantitative data is collected first and analysed, followed by the collection and analysis of qualitative data to help to explain the results of the quantitative data (Creswell & Hirose, 2019; Guest & Fleming, 2014). This design gives priority to the quantitative methods. This typology is most useful when the researcher wants to obtain in-depth insight to help explain the initial findings of the quantitative study (Harrison & Reilly, 2011).

For exploratory and explanatory concurrent mixed methods designs, the two forms of data (qualitative and quantitative) are collected and analysed simultaneously (Morse & Niehaus, 2016; Creswell, 2009). The researcher may decide to give priority to either the qualitative or quantitative data as a main component (Morse & Niehaus, 2016). The data collection and analysis processes of both research approaches are independent of each other (qualitative or quantitative) (Guest & Fleming, 2014). Although data collection is carried out simultaneously, priority can be given to the qualitative data or the quantitative data, depending on the objectives of the study. Similarly, equal priority can also be assigned to both forms of qualitative and quantitative data (Hanson *et al.*, 2005). The concurrent, mixed methods design is useful when the researcher wants to "confirm, cross-validate, and corroborate the findings of the study" involving the same phenomenon (Hanson *et al.*, 2005: 229).

For the embedded or nested mixed methods design, both the quantitative and qualitative data are collected and analysed either within a qualitative or quantitative design (Guest & Fleming, 2014). In the embedded design, the study is guided by primary data and supported by supplementary data, providing a supporting role in the research process (Creswell, 2009). Data can be collected sequentially or concurrently, with one of the datasets acting as the main component and the other playing a supplementary role within the larger design of the study (Hanson *et al.*, 2005). In concurrent embedded design, both qualitative and quantitative data are collected and analysed simultaneously. Similarly, for sequential embedded design, both forms of data are collected and analysed sequentially. Priority is usually given to the main components (data) in which the other supplementary component (data) is embedded. The embedded or nested design enables one form of data (e.g. qualitative) to be embedded or implanted within another form of data (e.g. quantitative) or vice-versa (Klingner & Boardman, 2011). This research design is useful when the researcher is interested in gaining a broader insight into a phenomenon and investigating different phenomena "within a single study" (Hanson *et al.*, 2005: 229). It is also suitable in a situation where there are different research questions which require different data to address the questions (Harrison & Reilly, 2011).

Motivation for Using Mixed Methods Research Designs

There are various reasons why researchers would want to combine multiple research methods to investigate a phenomenon. The motivation is primarily either to confirm, complement, or corroborate the findings from one method with another (Harrison & Reilly, 2011; Small, 2011). The principle underlying the motivation for confirmation is that researchers tend to have confidence in their findings when the findings of one method agree or align with the findings of the other method used for the investigation of the same phenomenon in a single study (Small, 2011). The principle of confirmation is useful when researchers are seeking to validate the findings by using the findings of one dataset to confirm or refute the findings of another dataset used to investigate the same phenomenon (Small, 2011). It is most suitable for concurrent or embedded designs.

The motivation for complementarity occurs when researchers want to use the findings from one research method to enhance or explain the findings of another research approach (Guest, 2012). The approach is deemed to be useful when researchers are "seeking elaboration, enhancement, illustration, and clarification of the results from one method with results from the other method" (Johnson & Onwuegbuzie, 2004: 22). For the purpose of complementarity, researchers can use the findings of one dataset (e.g. qualitative) to expand or explain the findings of another dataset (e.g. quantitative) in the investigation of the same phenomenon within a single study (McCrudden *et al.*, 2021). The findings of both datasets play complementary roles towards the understanding of the same phenomenon being investigated (McCrudden *et al.*, 2021). This approach is suitable for studies involving sequential, concurrent and embedded designs.

The motivation for corroboration occurs when researchers want to use the findings from one dataset to corroborate the interpretation of the findings of another dataset studying the same phenomenon in a single study. It is most appropriate for a concurrent design where researchers are seeking to ascertain corroboration or agreement between the findings of both datasets (e.g. qualitative and quantitative investigations carried out separately) about the same phenomenon in a single study (Wisdom *et al.*, 2012; Harrison & Reilly, 2011; Johnson & Onwuegbuzie, 2004).

Integrating and Merging Stage

For a mixed methods research design, there must be "at least one connecting point of integration" of the datasets (Doyle *et al.*, 2016: 631). Integration refers to the stage at which the datasets of two different research methodologies are brought together (Halcomb, 2019). It forms an important aspect of mixed methods research as it provides the opportunity for researchers either to corroborate, complement or confirm the findings from their studies. According to Harrison and Reilly (2011: 20), any mixed methods study in which two forms of datasets are used without integration is considered to be a "mere collection of methods". The use of the mixed methods research design requires a researcher to decide on how and where the integration of the datasets should take place (Schoonenboom & Johnson, 2017). There are many ways in which data from two different research approaches can be integrated during the research process (Bazeley, 2012). Data integration can occur at various levels/stages of the study. It can occur at design, data collection and analysis, and interpretation stages or at all the stages of the research process (Shannon-Baker, 2016; Creswell, 2009) sequentially and concurrently, depending on the typology of mixed methods design adopted and the objective of the study (Guest & Fleming, 2014). Figure 1.1 illustrates the levels of data integration for a mixed methods research process.

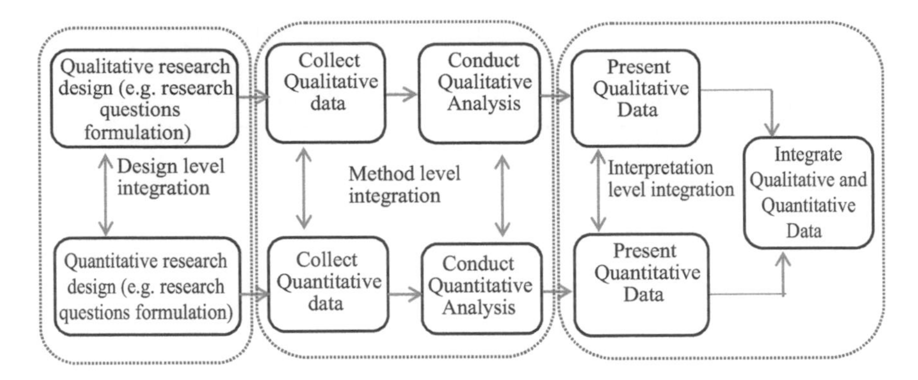

Figure 1.1 Levels of data integration for mixed methods research

Integration at the Design Level/Stage

Research questions requiring multiple research approaches to address them can be integrated at the design stage of the study (Doyle *et al.*, 2016). This is usually executed through the juxtaposition of these questions in a manner that suggests the utility of qualitative and quantitative data in articulating responses (Baškarada, & Koronios, 2018). According to these scholars, integration of such nature can be achieved at the design level using questions like "what and how"? wherein, the "what" question can be answered using quantitative data whilst the "how" question will rely on qualitative data (Baškarada & Koronios, 2018).

In the sequential exploratory and explanatory designs, the preceding method feeds into the subsequent method (Doyle *et al.*, 2016). For sequential exploratory mixed methods designs, the research questions of the initial, preceding qualitative design are used to inform the design and formulation of the research questions of the quantitative design. Likewise, the research questions of the preceding, quantitative design are used to formulate the research questions of the qualitative design in the sequential explanatory design process. Similarly, in the concurrent designs, the research questions for both methods (quantitative and qualitative) are designed almost at the same time during the study. For the embedded design, the integration of the research questions for both methods can occur simultaneously, or the research questions of the preceding method can be used to inform the formulation of the research questions for the supporting method (Harrison & Reilly, 2011).

Integration at the Method Level/Stage

At the method level (depending on the mixed methods design typology adopted), the integration occurs at various stages of the study. In sequential designs (exploratory and explanatory), the integration can occur when results from the preceding method are used to inform (connect) the data collection and analysis of the subsequent method (Doyle *et al.*, 2016; Guest & Fleming, 2014; Creswell, 2009). For sequential exploratory design, integration occurs when the findings from the initial, qualitative study is used to inform the data collection and analysis of the quantitative data. Similarly, for sequential explanatory design, integration takes place when the results from the initial quantitative study is used to inform the collection and analysis of the qualitative data (Doyle *et al.*, 2016).

However, data integration can also occur at the interpretation levels/stages in studies involving sequential designs. This approach to integration, which is akin to the concurrent design, has become an alternative approach adopted by many researchers. The best practice is to present findings and interpretations in the sequential order of the design (Fetters & Freshwater, 2015). For example, in exploratory sequential design, the qualitative findings are presented followed by the quantitative results (Fetters & Freshwater, 2015).

In concurrent designs, the two forms of datasets are first analysed separately, and the point of integration occurs where the two datasets are merged for comparison and interpretation. The data integration approach adopted for embedded designs is similar to the concurrent designs (when two forms of data are collected and analysed simultaneously), except where the preceding data is used to inform the collection and analysis of the supporting data for the overall study.

Integration at the Interpretation Level/Stage

Integration of data occurs when two forms of datasets are merged for comparison and interpretation (Fetters *et al.*, 2013). According to Plano Clark *et al.* (2010: 161), the integration of two forms of data at the final point of the study helps "to develop a more complete picture by presenting two complementary sets of results about a topic". The commonly used integration techniques for reporting findings at the interpretation stage include narrative, data transformation, and joint displays (McCrudden *et al.*, 2021; Moseholm & Fetters, 2017; Fetters & Freshwater, 2015).

Narrative

The commonly used narratives are weaving and contiguous techniques. The weaving narrative technique enables the findings of the two forms of datasets to be merged "together on a theme-by-theme or concept-by-concept basis" within a single report (Fetters *et al.*, 2013: 2142). The weaving helps researchers to integrate the findings of the two datasets (numerical and textual) thematically and/or conceptually into each other in a logical manner for comparison (Fetters *et al.*, 2013) and for "evidence of convergence, divergence, or contradiction between the two datasets" (Guest & Fleming, 2014: 594). The contiguous narrative technique, on the other hand, is used to present the findings of the two datasets separately in different sections in a successive manner within a single report (Fetters & Freshwater, 2015; Fetters *et al.*, 2013). Thus, the narrative of qualitative findings can be presented first, followed by the presentation of the narrative of quantitative results (and vice-versa, depending on the topology adopted) in a single report.

Data Transformation

In this reporting technique, one type of data, for example, qualitative data, can be transformed – "quantitised" – into quantitative data or quantitative data can be "qualitised" into qualitative data and then both data (qualitative and quantitative) can be integrated for comparison and interpretation of the findings (Moseholm & Fetters, 2017; Bazeley, 2012). Qualitising data means converting the quantitative data (numerical data) into textual data or words in consonance with the qualitative textual data for comparative analysis and interpretation of the findings. Conversely, quantitised data emerges when qualitative data (textual data) is converted into numerical data for comparison and interpretation of the findings (Bazeley, 2012).

Joint Display

The presentation of data by joint display enables findings from the two research methods (quantitative and qualitative) to be presented side by side for analysis and interpretation of the two datasets (McCrudden *et al.*, 2021; Moseholm & Fetters, 2017). Joint displays can be done by using visual forms such as tables, matrices, figures, graphs, or combinations of different visual forms (McCrudden *et al.*, 2021; Fetters *et al.*, 2013). This technique provides an opportunity for researchers to compare the datasets and identify relationships between and within the two datasets (McCrudden *et al.*, 2021). It is most useful for concurrent mixed methods designs. But it can also be used to present data for studies involving sequential designs (McCrudden *et al.*, 2021). For example, in sequential exploratory design, the results can be presented in columns in a table, first by reporting the findings of the qualitative study in the first column, followed by the results of the quantitative study in the second column, and, in the third column, "the impact of the integration in the study" is recorded (Creswell & Hirose, 2019: 6).

How the Book Is Structured

Overview of the Chapters

The book is made up of 17 chapters which are structured into four sections. Section 1 contains four chapters about the introduction, philosophical stance, ethical considerations, and a review of mixed methods research design in construction education. Section 2 contains six chapters on explanatory sequential mixed methods research design typology. Section 3 is made up of four chapters on exploratory sequential mixed methods research design typology, while Section 4 contains three chapters on convergent, embedded, and adaptive mixed methods research design typology. A brief insight into the contents of these chapters is presented below.

In Chapter 1, Introduction to Mixed Methods Research Designs in the Built Environment, Julius Akotia, Bankole Osita Awuzie, and Charles Egbu discuss a variety of mixed methods research design typologies, highlighting the advantages and disadvantages of various typologies. Furthermore, it presents the principle and theory behind the application of mixed methods research design in built environment research. It allows readers to develop a broader understanding of mixed research methods designs, the motivation for their adoption, and the processes associated with data integration in mixed methods research design. It forms the background for the other chapters.

In Chapter 2, Onto-Epistemological Assumptions Underpinning Mixed Methods Research Designs, Julius Akotia, Bankole Osita Awuzie, and Charles Egbu present a discussion of the philosophical principles underpinning the mixed methods research design. The chapter highlights the importance of philosophical assumptions and how such philosophical assumptions inform the selection of the research approach in line with the mixed methods research design.

In Chapter 3, Ethical Considerations in Mixed Methods Research Design, Abid Hasan discusses the ethical considerations that must be complied with when using a mixed methods research design in the context of the built environment research. Specifically, the ethical considerations that apply to different stages of a mixed methods research study. The chapter highlights that the lack of oversight or diligence in exercising ethical duties when undertaking mixed methods research could have profound moral, professional, financial, and legal implications for study participants, researchers, institutions, other stakeholders, and the outcomes of the study. The chapter further indicates that the identification of ethical issues and best practices to overcome ethical issues would enable built environment researchers to navigate the ethical challenges and hasten the ethics approval process, hence saving considerable time and resources.

In Chapter 4, A Review of Mixed Methods Research Design in Construction Education, Saeed Rokooei reviews literature relating to the use of mixed methods research design in construction education as published in a major construction journal over the past decade. This study was predicated on the need to explore the extent to which the knowledge generated through mixed methods research considers the situational nature of educational methods in the three areas of archival studies, empirical research, and simulation models. The chapter maps out deficiencies and limitations in these studies with regards to mixed methods research design features and highlights the pitfalls that are overlooked and thereby compromising the validity and reliability of these studies. In this chapter, the author also provides a methodological structure, strategies, and guidelines for developing experimental frameworks, data collection methods, modelling mixed method approaches for construction education research, and a roadmap for designing different types of mixed methods research design approaches for construction education studies.

In Chapter 5, A Hybrid Project Management Model for Construction Projects: A Mixed Research Approach, Abdallah Lalmi, Gabriela Fernandes, and Souad Sassi Boudemagh provide insights into the development of a hybrid project management model, which combines traditional project management with agile and lean approaches for the construction industry, using an explanatory sequential mixed methods research design. The contents of the chapter focus on the conceptualisation of a hybrid project management model based on a case study of a large-scale construction project of an Algerian tram system which construction organisations could adapt in different contexts to manage and deliver large-scale construction projects. Furthermore, the chapter demonstrates the utility of the mixed methods research design for studies seeking to conceptualise models like the Hybrid Management Model.

In Chapter 6, Enhancing the Employability of Quantity Surveying Graduates: A Mixed Methods Approach, Samuel Adekunle, Iniobong Beauty John, Obuks Ejohwohmu, Clinton Aigbavboa, Andrew Ebekozien, and Douglas Aghimien demonstrate the usefulness of the sequential explanatory mixed methods research design approach for investigating the disparity between the appropriate competencies required by employers and the ones that are acquired by QS graduates from higher education institutions. Based on its findings, the chapter proposes the realignment of extant curriculum by such institutions to ensure a perfect match between the competencies and skills required and acquired by employers and quantity surveying graduates, respectively, thereby bridging any gaps experienced.

In Chapter 7, Understanding the Effects of the Built Environment on Autistic Adults, Beth Noble and Nigel Isaacs expatiate on the utility of an explanatory sequential mixed methods research design for investigating the experience of autistic and neuro-typical people as it pertains to certain built environment features within the New Zealand context. Particularly, the study elicits the response of constituents of this population, to various aspects of the built environment Indoor Environment Quality (IEQ) factors. In the first phase of the study, data is collected using a questionnaire survey which was administered to participants through social media, local businesses, word-of-mouth, disability advisory organisations, and local autistic-led self-advocacy organisations. In the second phase, a qualitative participatory photographic approach is used to enable autistic people to self-select examples of indoor electric lighting that they like or dislike by sending photographs and descriptions about the reasons for their choices. Summarily, the chapter provides guidelines on how to deploy the mixed methods research design for understudying physiological effects of the built environment on distinct populations.

In Chapter 8, A Mixed Methods Evaluation of the Social Value of Indigenous Procurement Policies in the Australian Construction Industry, George Denny-Smith, Riza Yosia Sunindijo, Megan Williams, Martin Loosemore, and Leanne Piggott articulate the use of explanatory sequential mixed methods research design in studying the social value created through Australian Indigenous Procurement Policies (IPPs) relying on the Ngaa-bi-nya aboriginal evaluation framework. The chapter presentes a practical guide on how mixed methods research designs could be applied by built environment researchers involved with community-based evaluation studies to gain conceptual and practical insights by outlining the critical decisions and methodological options available to perform high-quality research that explains novel social phenomena.

In Chapter 9, A Methodological Application to Construction Economics Research for Theory Refinement and Extension, Mohd Azrai Azman, and Carol Hon understudy the diversification and productivity performance of large construction firms in Malaysia using an explanatory sequential mixed methods research design. The chapter outlines how mixed methods research design can be used to support the results from econometric models to provide a meaningful explanation for causes and effects using theoretical and practical perspectives. The mixed methods research design was used to validate the quantitative results and to refine the economic theory regarding the diversification of construction firms in the context of Malaysian construction firms. The chapter serves as a guide for researchers to apply mixed methods research design in construction economics research with prospects for theory refinement and extension.

In Chapter 10, Effects of Physical Infrastructure on Practical Performance of Graduates in Architectural Technology in Southeast Nigeria: An Explanatory Sequential Mixed Methods Research Design, Udochukwu Marcel-Okafor outlines the practicality of using explanatory sequential mixed methods research design to assess the impact of physical infrastructure facilities on the ability of architectural technology graduates to achieve the stipulated learning outcomes and achieve the relevant competences associated with the programme. The author used a mix of structured questionnaires and interviews for the elicitation of quantitative and qualitative data within a multiple case study context. The chapter demonstrates the usefulness of mixed methods research designs in engendering effective, comprehensive, and inclusive understanding of the relevant phenomenon within the field of architectural technology education.

In Chapter 11, An Exploration of Sustainable Procurement Practice in Irish Construction-Contracting Firms, Duga Ewuga, Mark Mulville, and Alan Hore employ an exploratory sequential mixed methods design approach to explore and assess the significance of organisational strategies adopted by the top 50 large construction-contracting firms in the Republic of Ireland in driving sustainable procurement practice implementation. Throughout the chapter, the authors illustrate how the use of mixed methods research design enabled the collection of a rich and robust array of evidence to explore extant organisational strategies, policies, practices, and performances of large construction-contracting firms in the Republic of Ireland as it pertains to sustainable procurement practice. Furthermore, the utility of the design in enabling the identification of areas requiring improvement is elucidated in the chapter.

In Chapter 12, Investigating the Accident Causal Influence of Construction Project Features Using Sequential Exploratory Mixed Methods Design, Patrick Manu illustrates the use of a sequential exploratory mixed methods approach in examining the accident causal influence of construction project features and the extent of their impact on health and safety. The chapter offers insights into the rational process that leads to the selection of a mixed methods approach and, subsequently, the steps taken to operationalise the approach to address the research questions. It serves as a useful reference to guide other health and safety researchers in the selection and deployment of the sequential exploratory mixed methods research design to address similar or different research questions in construction accident causation issues.

In Chapter 13, Infrastructure Project Selection and Prioritisation for Socio-economic Development in Mining Communities of Ghana: A Sequential Mixed Methods Research Approach, Enoch Sackey, Julius Akotia, and Jemima Ottou adopt a sequential exploratory mixed methods research approach to explore the selection and prioritisation criteria for infrastructure projects for optimising the harnessing of scarce resources and enhancement of socio-economic development in Ghana's deprived mining communities. The chapter elucidates the relevance of the adopted research design in deciphering the subconscious biases that affect decision-makers during the formulation of salient project selection and prioritisation criteria and appropriate strategies for mitigating such biases. It is expected that such understanding would facilitate the achievement of a consensus on relevant criteria to facilitate development in communities with skewed infrastructure services and inadequate socio-economic capacity.

In Chapter 14, Exploratory Sequential Mixed Method Research to Investigate Factors Affecting the Reputation of PFI/PF2 Projects in the United Kingdom, Stanley Njuangang, Henry Abanda, Champika Liyanage, and Chris Pye deploy an exploratory sequential mixed methods research design to investigate the factors that affect the reputation of PFI projects in the United Kingdom. The grounded theory (GT) approach is used as the primary method for collecting qualitative data, whereas a questionnaire survey is used for eliciting quantitative data. The chapter outlines the processes involved in the selection of the documents for carrying out the GT, the elicitation of PFI reputational issues from literature, and how quantitative data was subsequently collected using a questionnaire survey. Furthermore, the chapter illustrates the integration of qualitative and quantitative data using joint displays to bring about more meaningful interpretation of the findings.

In Chapter 15, An Exploration of the Implications of Sustainable Construction Practice: Mixed Methods Research Approach, Cheng Siew Goh and Shamy Yi Min Chin demonstrate the utility of the concurrent mixed methods research design for investigating stakeholder perceptions as it concerns the comprehensive implementation of sustainable construction practices. The use of the concurrent mixed methods research design enabled an appreciation of the extant realities associated with the delivery of sustainable construction whilst uncovering the extent to which each of the three pillars of sustainability – environmental, social, and economic are implemented by stakeholders in their quest for optimal sustainable construction practice. Summarily, the chapter provides a succinct illustration of the integration of qualitative and quantitative data.

In Chapter 16, Using Convergent Mixed Method to Explore the Use of Recycled Plastics as an Aggregate for Concrete Production in South Africa, Nishani Harinarain, Robert Lukan, and Bevan Naidoo adopte a convergent mixed methods research design to investigate the willingness of construction industry professionals to use recycled plastics as a substitute for sand as an aggregate in concrete production. The collection of quantitative data (questionnaire survey) and qualitative data from the built-environment professionals is done concurrently. Also, the results from the survey are integrated with the findings from the interviews in a logical and coherent manner. The use of a convergent mixed methods design enabled the authors to obtain a broader and richer perspective on the topic and increased the credibility of the findings.

In Chapter 17, Adaptive Mixed Methods Research for Evaluating Community Resilience and the Built Environment, Sandra Carrasco and Temitope Egbelakin demonstrate the practicality and flexibility of an adaptive mixed methods research design for understudying community resilience and its implications for creating a safer built environment using two case studies situated in the Philippines and New Zealand, respectively. The successful seamless implementation of various data collection instruments and approaches across two case studies further accentuated the need for methodological flexibility and adaptability when conducting investigations into disaster-related problems in the built environment context. The use of the adaptive mixed methods research design enabled the authors to cater to the permanently changing conditions of disaster recovery and preparedness as illustrated in the chapter. The features associated with this research design provide opportunities for incorporating novel approaches and techniques in conducting studies within such contexts and knowledge domains.

References

Amaratunga, D., Baldry, D., Sarshar, M. & Newton, R. 2002. Quantitative and qualitative research in the built environment: Application of "mixed" research approach. *Work Study*, 51(1): 17–31.

Baškarada, S. & Koronios, A. 2018. A philosophical discussion of qualitative, quantitative, and mixed methods research in social science. *Qualitative Research Journal*, 18(1): 2–21.

Bazeley, P. 2012. Integrative analysis strategies for mixed data sources. *American Behavioral Scientist*, 56(6): 814–828.

Brown, L. 2021. A philosophy, a methodology, and a gender identity: Bringing pragmatism and mixed methods research to transformative non-binary focused socio-phonetic research. *Toronto Working Papers in Linguistics (TWPL)*, 43. https://doi.org/10.33137/twpl.v43i1.35965

Bryman, A. 2006. Paradigm peace and the implications for quality. *International Journal of Social Research Methodology*, 9(2): 111–126.

Chen, H.T. 2006. A theory-driven evaluation perspective on mixed methods research. *Research in the Schools*, 13(1): 75–83.

Combs, J.P. & Onwuegbuzie, A.J. 2010. Describing and illustrating data analysis in mixed research. *International Journal of Education*, 2(2): E13.

Creswell, J.W. 2009. *Research Design: Qualitative, Quantitative and Mixed Methods Approaches*. 3rd edition. Thousand Oaks, CA: Sage Publications.

Creswell, J.W. & Garret, A.L. 2008. The "movement" of mixed methods research and the role of educators. *South African Journal of Education*, 28: 321–333.

Creswell, J.W. & Hirose, M. 2019. Mixed methods and survey research in family medicine and community health. *Family Medicine and Community Health*, 7.

Day, J.K. & Gunderson, D.E. 2018. Mixed methods in built environment research. *54th ASC Annual International Conference Proceedings, Texas*, 1–6, http//www.ascpro.ascweb.org.

Doyle, L., Brady, A.M. & Byrne, G. 2016. An overview of mixed methods research. *Journal of Research in Nursing*, 21(8): 623–635.

Fetters, M.D., Curry, L.A. & Creswell, J.W. 2013. Achieving integration in mixed methods designs – Principles and practices. *Health Services Research*, 48(Part II): 6.

Fetters, M.D. & Freshwater, D. 2015. Publishing a methodological mixed methods re-search article. *Journal of Mixed Methods Research*, 9(3): 203–213.

Greene, J.C. 2008. Is mixed methods social inquiry a distinctive methodology? *Journal of Mixed Methods Research*, 2(1): 7–22.

Grix, J. 2004. *The Foundations of Research*. Palgrave, MacMillan, London.

Guest, G. 2012. Describing mixed methods research: An alternative to typologies. *Journal of Mixed Methods Research*, 7(2): 141–151.

Guest, G. & Fleming, P.J. 2014. Mixed methods research. In: G. Guest & E. Namey (eds). *Public Health Research Methods*, pp. 581–610. Sage, Thousand Oaks, CA.

Halcomb, E.J. 2019. Mixed methods research: The issues beyond combining methods. *Journal of Advance Nursing*, 75: 499–501.

Hanson, W.E., Creswell, J.W., Plano Clark, V.L., Petska, K.S. & Creswell, J.D. 2005. Mixed methods research designs in counseling psychology. *Journal of Counseling Psychology*, 52(2): 224–235.

Harrison, R.L. & Reilly, T.M. 2011. Mixed methods designs in marketing research. *Qualitative Market Research: An International Journal*, 14(1): 7–26. Johnson, B. & Onwuegbuzie, A.J. 2004. Mixed methods research: A research paradigm whose time has come. *Educational Researcher*, 33(7): 14–26.

Johnson, R.B., Onwuegbuzie, A.J. & Turner, L.A. 2007. Toward a definition of mixed methods research. *Journal of Mixed Methods Research*, 1(2): 112–133.

Klingner, J.K. & Boardman, A.G. 2011. Addressing the "Research Gap" in special education through mixed methods. *Learning Disability Quarterly*, 34(3): 208–218.

Masadeh, M.A. 2012. Linking philosophy, methodology, and methods: Toward mixed model design in the hospitality industry. *European Journal of Social Sciences*, 28(1): 128–137.

Maxwell, J.A. 2016. Expanding the history and range of mixed methods research. *Journal of Mixed Methods Research*, 10(1): 20, 12–27.

McCrudden, M.T., Marchand, G. & Schutz, P.A. 2021. Joint displays for mixed methods research in psychology. *Methods in Psychology*, 5, 100067.

Mingers, J. 2001. Combining is research methods: Towards a pluralist methodology. *Information Systems Research*, 12(3): 240–259.

Morse, J.M. & Niehaus, L. 2016. *Mixed Method Design: Principles and Procedures*. Routledge, New York.

Moseholm, E. & Fetters, M.D. 2017. Conceptual models to guide integration during analysis in convergent mixed methods studies. *Methodological Innovations*, 10(2): 1–11.

Newman, I. & Hitchcock, J.H. 2011. Underlying agreements between quantitative and qualitative research: The short and tall of it all. *Human Resource Development Review*, 10(4): 381–398.

Onwuegbuzie, A.J. & Johnson, R.B. 2006. The validity issue in mixed research. *Research in the Schools*, 13(1): 48–63.

Panas, A. & Pantouvakis, J.P. 2010. Evaluating research methodology in construction productivity studies. *The Built & Human Environment Review*, 3(Special Issue 1): 63–85.

Plano Clark, V.L. 2010. The adoption and practice of mixed methods: U.S. trends in federally funded health-related research. *Qualitative Inquiry*, 16(6): 428–440.

Plano Clark, V.L., Garrett, A.L. & Leslie-Pelecky, D.L. 2010. Applying three strategies for integrating quantitative and qualitative databases in a mixed methods study of a nontraditional graduate education program. *Field Methods*, 22(2): 154–174.

Saunders, M., Lewis, P. & Thornhill, A. 2009. *Research Methods for Business Students*. 5th edition. London: Pearson Education.

Schoonenboom, J. & Johnson, R.B. 2017. How to construct a mixed methods research design. *Köln Z Soziol*, 69(Suppl 2): 107–131.

Shannon-Baker, P. 2016. Making paradigms meaningful in mixed methods research. *Journal of Mixed Methods Research*, 10(4): 319–334.

Small, M.L. 2011. How to conduct a mixed methods study: Recent trends in a rapidly growing literature. *Annual Review of Sociology*, 37: 57–86.

Tashakkori, A. & Teddlie, C. 2010. Putting the human back in "Human Research Methodology": The researcher in mixed methods research. *Journal of Mixed Methods Research*, 4: 271.

Wisdom, J.P., Cavaleri, M.A., Onwuegbuzie, A.J. & Green, C.A. 2012. Methodological reporting in qualitative, quantitative, and mixed methods health services research articles. *Mixed Methods in Health Services Research*, 47: 2.

2 Onto-Epistemological Assumptions Underpinning Mixed Methods Research Designs

Julius Akotia, Bankole Osita Awuzie, and Charles Egbu

Summary

Philosophical assumptions play a significant role in the selection of methods for executing research projects. The relevance of such assumptions in the selection of appropriate mixed methods research design typologies for research projects in the built environment discipline does not fare any differently. Predicating the choice of mixed method research design to be deployed for a study on an appropriate philosophical stance or paradigm enables researchers to reflect on practical ways to view and construct the nature of reality, discover knowledge and address ethical or values issues associated with their research. This chapter presents a discussion of the philosophical assumptions underpinning the mixed methods research design. It highlights the importance of philosophical assumptions and how philosophical assumptions inform the selection of the mixed methods research design. It presents discussion on pragmatic and critical realist philosophies and how they are aligned with the mixed methods research design.

Research Philosophy

Burke (2007: 476) defined research philosophy "as the questioning of basic fundamental concepts and the need to embrace a meaningful understanding of a particular field". Burke (2007) noted that the pursuit of such a philosophical perspective provides a useful starting point, as it makes it possible to communicate the research approach clearly in a context that is well understood by others. The exploration of philosophical assumptions is a crucial aspect in establishing a researcher's philosophical outlook, which ultimately helps to clarify the research approach, leading to the development of a suitable research methodology, methods of data collection and the method of analysis. Similarly, philosophical assumptions play a key role in influencing and directing the way knowledge and meaning in a particular field can best be captured and interpreted (Burke, 2007). Saunders *et al.* (2009) suggested that the philosophical propositions made by a researcher through the selection of a research strategy have a considerable impact, not only on what the researcher does but also on how the researcher understands what it is that is being investigated. According to Morgan (2007), and Kaushik and Walsh (2019), philosophically informed research

DOI: 10.1201/9781003204046-2

plays a significant role, as shared beliefs form the basis for the selection of a research paradigm, the data collection strategy and method of analysis. Moreover, the philosophical assumptions that underlie a selected paradigm adopted to study a particular phenomenon will enable the researcher to be aware of the boundaries within which to approach the investigation. Accordingly, understanding philosophical assumptions of a research area provides the necessary knowledge for the researcher to situate the research strategy within a suitable research paradigm in relation to the researcher's belief systems (Burke, 2007).

Seeking to gain an understanding of what is perceived as valid knowledge requires the exploration of such philosophical assumptions as essential aspects of the research process to ensure that the research is conducted with rigour and with credibility (Burke, 2007). However, achieving such objectives requires that the research is established within a suitable paradigm, with clearly stated philosophical assumptions relevant to that research paradigm (Burke, 2007). The discovery of knowledge is the epitome of any research endeavour. For that reason, the pursuance of knowledge generation must be well grounded in well-defined philosophical assumptions (Ardalan, 2009). Brennan *et al.* (2011) viewed the creation of knowledge as a pre-requisite of a research project, which can only be achieved through a solid understanding and application of philosophical assumptions. Saunders *et al.* (2009) identified a link between philosophical assumptions and the creation of knowledge in relation to research approaches adopted to investigate a phenomenon. According to Adcroft and Willis (2008), the philosophical assumptions, whether implicitly or explicitly expressed, provide the logical starting point for endeavours in knowledge enquiry.

The philosophical assumptions also have a significant influence on the methodological approaches of research and the processes and methods of data collection adopted for the research. Brennan *et al.* (2011: 103) believed that

> any approach to knowledge inquiry rests upon certain foundational assumptions and presuppositions, about the nature of reality, about the nature of possible forms of knowledge about that reality, about the types of methods which can be used to generate that knowledge, and several others.

It is argued that different research paradigms lend themselves to different forms of assumptions and propositions to knowledge enquiry, and these assumptions and propositions tend to identify with specific philosophical elements, appropriate to such research paradigms (Brennan *et al.*, 2011). According to Brennan *et al.* (2011: 103), such an "approach to knowledge inquiry (or paradigm) inspires certain commitments and assumptions, which are inherent in the paradigm – including ontological, epistemological, axiological, and methodological" aspects of research philosophy. Thus, the epistemological aspect provides the link between practical and theoretical knowledge, strengthened by ontological and axiological considerations, as fundamental facets of research enterprise (Carter & Little, 2007). These considerations determine how researchers view and conceptualise reality (ontology), knowledge generation (epistemology) and address ethical or value issues (axiology), which provide a guiding framework for the consideration of the methodology, as well as the methods of data collection and analysis. The ontological underpinnings determine the epistemological and axiological positions of the researcher, and these positions ultimately determine the choice of the methodology and the data collection (Mack, 2010). Brennan *et al.* (2011: 107) believed that the philosophical assumptions are "contingent on both the inputs to the research as well as the procedures undertaken within the research process itself, for any research to be considered valid and reliable". Philosophical assumptions enable the researcher to reflect on practical ways to view and construct the nature of reality (ontology), discover knowledge (epistemology) and address ethical or values issues (axiology) for the research.

In their earlier work, Guba and Lincoln (1988) identified four philosophical assumptions of research paradigms (positivist and interpretivist) in terms of their epistemology (how researchers know what they know), ontology (the nature of reality), axiology (the place of values in research) and methodology (the process of research). According to Ruona and Lynham (2004), philosophically grounded research lends itself to the interaction of three assumptions: epistemology (which makes claims about the nature of knowledge), ontology which is concerned with the fundamental assumptions about the nature of reality, and axiology (which is concerned with the ethical or value issues that guide the research). In the view of Denzin and Lincoln (2005), the set of belief systems that guide research action, which also reflects the researcher's worldview and perceptions, is composed of four sets of philosophical assumptions: epistemology (knowledge), ontology (reality), axiology (ethics or morality) and methodology (enquiry). Carter and Little (2007) agreed to the above position by adding that making a valid assumption philosophically has the propensity for a viable research outcome because it influences the selection of the research methodology and the methodological processes of the research. In that sense, a careful selection of the research philosophy, methodology and methods will enable researchers to address their research objectives (Carter & Little, 2007). The researcher's philosophical underpinnings are fundamental in determining the "research design, the main research questions and subsequently the chosen methodology" (Almpanis, 2016: 302).

As such, making such philosophical assumptions in mixed methods research remains of paramount importance to the selection of the research design that transcends epistemological and ontological barriers by combining either or a combination of inductive, deductive, abductive and retroductive research inferences. In line with the philosophical underpinnings, the researcher is able to select the research methodology and data collection methods which are consistent with the elements of mixed methods research. Morgan (2007) intimated that a strong appreciation and application of such philosophical assumptions is crucial in discovering the knowledge relevant to the use of a mixed methods research approach. Scholars have highlighted the difficulty of situating the mixed method research design as a paradigm within a particular philosophical stance (Zou *et al.*, 2014). This difficulty has been attributed to the differences experienced in the epistemological inclinations of qualitative and quantitative methods of data collection and analysis.

Corroborating previously espoused views, Baškarada and Koronios (2018), highlight the salient contribution of appropriate philosophical stances in engendering the logical determination and achievement of relevant criteria for measuring the validity of the study's research instruments and findings. The authors argue that, in the absence of such a logically determined validity construct, the attainment of such criteria, an important contributor to the credibility of the study's outputs, would be rather subjective (Baškarada & Koronios, 2018). The quest to achieve this credibility criteria has been deduced to pose a challenge to mixed methods research-led studies, particularly in the built environment. Perhaps, this has led to the increasing reliance of built environment researchers on pragmatism as a philosophical stance of choice (Alise & Teddlie, 2010).

Venkatesh *et al.* (2016) categorised philosophical stances which are available for facilitating mixed methods research into dialectic, alternative and complementary strength stances categories. According to these authors, whereas the dialectic category allowed for the use of a combination of research paradigms on the same research project to engender a more in-depth insight into a phenomenon, the alternative paradigm category consists of pragmatism and critical realism philosophies (Venkatesh *et al.*, 2016). Furthermore, the complementary strengths stance indicates the possibility of incorporating other philosophical stances usually associated with social science and humanities disciplines in designing mixed methods research projects (Venkatesh *et al.*, 2016). Similarly, Shan (2022) in a contribution to the philosophical stance debate in mixed methods research, delineated these paradigms into the following positions, namely, the pragmatist position, the dialectical position, the dialectical pluralist position, the transformative position and the critical realist position. These positions share similar meanings with the groupings provided by Venkatesh *et al.* (2016), especially as it concerns the dialectical and dialectical pluralist positions. Whilst the former was described as acknowledging the use of other philosophical stances like postpositivism to assist the researcher in articulating a variety of inquiry decisions in a particular study, the latter position is associated with a belief in the existence of multiple realities which can only be thoroughly explained or explored using different philosophical stances (Shan, 2022). Furthermore, this position, which is analogous to the complementary strength stance, appreciates the fallibility, contextual and value-laden nature of knowledge and as such allows for a combination of paradigms for such exploration. Although the transformative position emphasises the social construction of multiple realities, relying on norms and values among other social parameters, its main focus is on the use of mixed methods research to provide a truly representative perspective of the phenomenon being studied, leveraging the individual strengths of the methods deployed (Shan, 2022). In this sense, the position is associated with society's quest for social justice. Critical realism was identified as the last of the positions highlighted by Shan (2022). According to the author, this position which is predicated on the need to provide a qualitatively led understanding of causality whilst allowing for the use of quantitative methods to ascertain the magnitude of the causality and the contribution of the underlying mechanisms to the magnitude experienced has been described by scholars as having a potential to contribute to successful mixed method research (Shan, 2022).

Quantitative	Research Philosophy	Qualitative
Positivism	Ontology	Interpretivism
Objectivity	Epistemology	Subjectivity
Value Free	Axiology	Value laden
Quantitative	Methodology	Qualitative
Questionnaire	Method	Interview/observation

Figure 2.1 Philosophical underpinnings of mixed-methods research

Figure 2.1 illustrates the philosophical underpinnings of a mixed methods design involving quantitative and qualitative methodologies.

Judging from the plethora of mixed methods research designs utilised by built environment researchers, it can be discerned that pragmatism is most prevalent with critical realism posting a distant second. Therefore, subsequent sections provide detailed insight into both philosophies.

Pragmatic Philosophy

Pragmatism has been acknowledged as the most appropriate philosophical stance for investigating phenomena consisting of numerical and non-numerical elements (Kral *et al.*, 2012). As a philosophy, it is regarded as being the best fit for conducting research into complex human phenomena, by employing multiple research methodologies. A pragmatic philosophy assumes that there is only one, single world in which "individuals have their own unique interpretations of that world" and hence "treat issues of inter-subjectivity as a key element of social life" (Morgan, 2007: 72). Fundamentally, the philosophy embodies two distinct research paradigms: qualitative-interpretivism and quantitative-positivism. The pragmatic paradigm provides the flexibility for researchers to adopt multiple research methods with the objective of finding solutions to their research questions (Brierley, 2017). Pragmatists seek to capture inferences provided by both qualitative and quantitative methodologies into a single research framework to explore issues, particularly where they are of a multi-faceted nature (Fidel, 2008). Pragmatism does not ascribe or limit its philosophical orientations and prowess to one single world view: ontology (positivism or interpretivism), epistemology (objectivity or subjectivity), axiology (value-free or value-laden), methodology (quantitative or qualitative) and data collection method (questionnaire or interview) (Creswell, 2009). The central feature of a pragmatic philosophy is that the research approach can take both positivist and interpretivist ontological positions. Thus, the research approach can embrace both textual and measurable languages, which the researcher can use to investigate and understand the phenomenon under investigation (Morgan, 2007). Central to the philosophy of pragmatism is its ability to refute the paradigm debate that has existed between the qualitative-interpretivist and quantitative-positivist positions, by drawing on their respective strengths (Masadeh, 2012). According to Morgan (2007), such defining features of pragmatism provide the foundation for researchers to undertake investigations with what are traditionally regarded as being incompatible and conflicting paradigms. Pragmatic researchers are not aligned to one research ideology or world view. While a positivist philosophical orientation is aligned with quantitative methodology, an interpretivist philosophical underpinning is associated with qualitative methodology (Brown, 2021). This implies that researchers can draw from different research paradigms and methods to investigate complex and multi-faceted issues (Creswell, 2009).

Pansiri (2005) argued that the pragmatic philosophy provides a useful middle ground for combining different types of research approaches into a single workable framework, where the emphasis is on addressing practical research problems "that works" instead of focusing on the ideology of a particular research approach. A pragmatic philosophical orientation places research problems at the centre of the investigation. Drawing on the core values of the pragmatic orientation, Morgan (2007: 73) noted that it "offers an effective alternative through its emphasis on the abductive–intersubjective–transferable aspects of our research". According to Ardalan (2009: 516), such core values define the pragmatic research orientation in which both quantitative and qualitative approaches "share common fundamental assumptions about the nature of social science and the nature of society". Ardalan (2009) suggested further that different research paradigms are established based on different propositions about the nature of the scientific and social world, in which each of these paradigms generates their own assumptions and theories.

Various authors have identified pragmatism as the "philosophical partner" of mixed methods research (Pansiri, 2005: 201) that draws its strength from both the objective and subjective perspectives of knowledge generation. It is considered the appropriate philosophy and research paradigm that provides the best mechanism for researchers to answer inductive-based and deductive-based research questions all together within a single study. Similarly, the epistemological and ontological underpinnings inherent in the pragmatic research paradigm provide the philosophical basis and motivation for the choice of mixed methods research. Saunders *et al.* (2009: 598) believed that these philosophical underpinnings make the pragmatic paradigm an idealistic alternative according to which it is "possible to work within both positivist and interpretivist positions". Johnson and Onwuegbuzie (2004) are of the view that the essence of adopting the pragmatic research paradigm is that it offers an explicitly knowledge-oriented and practical approach to the examination of a phenomenon. More importantly, pragmatic research is focused on knowledge generation through what Morgan (2007: 72) referred to as "joint actions", which can be achieved together by multiple sets of research approaches. Pragmatic researchers are of the view that reality and knowledge of social phenomena can be obtained through the collaboration of "elements of both positivism and constructivism rather than solely one or the other" (Krauss, 2005: 767). Ardalan (2009) identified with the above positions by acknowledging the potency of a pragmatic research approach in the generation of knowledge which goes beyond the use of one research paradigm. Accordingly, adopting such a philosophical orientation facilitates the creation of knowledge in a manner that is consistent with the mixed-methods research approach.

While the traditional, individual, quantitative and qualitative methodologies emphasised deductive or inductive logic to knowledge generation of the social and scientific world, the pragmatic research paradigm goes a step further by providing a middle ground – an abductive approach (Morgan, 2007) and, by so doing, rejecting the barriers perceived to be between the traditional qualitative and quantitative-oriented methodologies (Masadeh, 2012; Shannon-Baker, 2016; Brierley, 2017). Although their respective usefulness is evident in different research contexts, pragmatism is largely considered to be the most appropriate and suitable research paradigm or philosophy that provides the best results for a mixed methods research approach (Tronvoll *et al.*, 2011). The philosophy of pragmatism is deeply rooted in mixed methods research where every assumption the researcher believes "to be real, including subjective realism, intersubjective realism, and objective realism" can be considered within a single research framework (Onwuegbuzie & Johnson, 2006: 54). For instance, Ardalan (2009) argued that much benefit can be achieved through the effective corroboration between the application of qualitative and quantitative research methodologies in a single study. Adopting the pragmatic research approach implies that the researcher is oriented towards seeking a dialogue between two distinct research approaches in achieving some particular research objective(s). It also means that the researcher is employing a research philosophy that is the best fit and oriented towards addressing multiple research problems (Brown, 2021). It has been suggested generally that the combination of the subjective and objective research paradigms in a single study provides a greater opportunity to investigate a complex phenomenon to enhance and expand the understanding of such a phenomenon (Morgan, 2007; Creswell, 2009).

The advantages of combining the perspective of both qualitative and quantitative research approaches despite the differences in their philosophical orientations have been acknowledged (Grix, 2004). Numerous questions have been raised about the fundamental issues relating to the philosophical orientations of mixed methods research. Previous contributors have sought to argue that the philosophical barriers between the two methods, coupled with their contrasting views, make the traditional, qualitative and quantitative methodologies incompatible to combine their elements. Moreover, according to Onwuegbuzie and Johnson (2006: 59), the processes involved in combining their elements have also been deemed to be tenuous because of the "competing dualisms: epistemological (e.g. objectivist vs subjectivist), ontological (e.g. single reality vs multiple realities), axiological (e.g. value free vs value-bound), methodological (e.g. deductive logic vs inductive logic), and rhetorical (e.g. formal vs informal writing style) beliefs" they espouse. However, while the two research methodologies seem to be espousing different philosophical ideologies, they tend to provide a research approach and philosophical dimension that seek to bring their perspectives together into a single practical solution (Johnson & Onwuegbuzie, 2004). The combination of such distinct philosophical underpinnings is the unique feature that defines mixed methods research. It can be argued that both qualitative and quantitative research methodologies have commonly acknowledged criteria that transcend these differences and barriers (Bryman, 2006). In practical terms, there are apparent overlaps to some extent between the two methodological ideologies. To a very large extent, this plays down the "difference" the argument perceived to be existing between them. For this reason, by de-emphasising their philosophical differences (Chen, 2006) and aiming solely at their potential benefits, the perceived differences between the two methodological ideologies are relegated to the background.

Critical Realism Philosophy

Although the foregoing highlights the prevalence of pragmatism in mixed methods approach-led built environment research, Venkatesh *et al.* (2016) maintain the utility of other philosophies in facilitating successful mixed methods research, particularly from an epistemological perspective. Of significance among the plethora of paradigms that can be used in mixed methods research is critical realism. This philosophical stance is positioned midway between positivism and interpretivism (Zachariadis *et al.*, 2013). Saunders *et al.* (2019) maintained that this philosophy is centred on providing an explanation of what an individual sees and experiences based on the contextual structures of reality which shape observable events. Schoonenboom (2019: 286) observed that the aim of the critical realism (CR) philosophical stance was to "obtain knowledge of a reality that is taken to exist in independent, objective way". Therefore, this philosophical stance would be appropriate for studies seeking to emphasise the experiences and feelings of individuals in trying to conceptualise their version of a given reality. It is suffice to state that critical realism underscores the existence of one reality which can be known from various perspectives (Schoonenboom, 2019). Such understanding extends legitimacy to the use of a multiplicity of methods to elicit perceptions of various individuals to engender the development of multiple realities relating to a particular phenomenon.

This philosophy is supported by the retroductive reasoning logic which ensures that an individual reasons backwards in terms of making sense of the underlying causes of an experience long after the events associated with that experience (Saunders *et al.*, 2019). Blaike (2007) maintained that the aim of the retroductive reasoning was to unravel the underlying mechanisms to explain observed regularities. To actualise this objective, an overt reliance on a juxtaposition of methods remains imperative, hence the increasing advocacy for the use of mixed methods research design in this regard (Maxwell & Mittapalli, 2010). Furthermore, Zachariadis *et al.* (2013) outlined a retroductive approach of critical realism to knowledge creation comprising three domains – the transitive, empirical knowledge and intransitive domains. The transitive domain consisted of contributions made by the iterative movement between quantitative magnitude and qualitative case enquiry of a phenomenon, which is based on old knowledge, as enabled by mixed methods research design towards achieving new knowledge development. This results in the empirical "knowledge" domain which further informs retroduction between actual events and real causes in the intransitive domain.

Critical realism has been used as a philosophical stance of choice in underpinning several mixed methods research in a variety of disciplines. Ryba *et al.* (2022) elucidated the relevance of critical realism in conducting mixed methods research projects in studies relating to sport and exercise psychology. In another study, Modell (2009) highlighted the salient contribution of critical realism in facilitating seamless triangulation in a mixed methods study focusing on management accounting. Mukumbang (2023) illustrated the importance of critical realism in providing a sound philosophical underpinning for mixed methods research study in a manner that allowed for effective retroductive reasoning in social sciences research. Also, Zachariadis *et al.* (2013) outlined the methodological implications associated with the adoption of critical realism in mixed methods research design albeit within the context of information system (IS) research. They highlighted the relevance of this philosophical stance in practice-oriented research with aspects relating to natural and social science knowledge domains respectively like the built environment domain.

In adopting a critical realism philosophy, the researcher is expected to apply the relevant epistemological paradigm towards establishing the process and conditions enabling a generative mechanism to result in an event or phenomenon and not the relationship between the mechanism and the event from a causality perspective (Zachariadis *et al.*, 2013; Venkatesh *et al.*, 2016). The stratified ontology of critical realism as proposed by Bhaskar (1975) as cited in Saunders *et al.* (2019) details the various strata of the CR paradigm highlighting an alignment between relevant dimensions at the domain level and the entity level. Such alignment starts from the "real" which consists of the prevailing properties of structures and mechanisms, the "actual" which focuses on events that are brought about by the presence of structures and mechanisms, whilst the "empirical" focuses on events which have been observed or experienced (Zachariadis *et al.*, 2013).

Various reasons have been deduced as necessitating the combination of methods (mixing of methods) in critical realism (Zachariadis *et al.*, 2013). These reasons consist of complementarity, completeness, developmental, expansion, corroboration, compensation and diversity, respectively. Critical realism just like pragmatism has posted considerable implications for the attainment of these purposes within the realm of mixed methods research (Shaw *et al.*, 2010; Zachariadis *et al.*, 2013; Brierley, 2017; Ryba *et al.*, 2022).

Shan (2022) classified the various philosophical stances into three distinct categories, this time depending on the strength of their philosophical foundations in making a case for the mixing of methods in contemporary research. These categories include the weak sense, the moderate sense and the strong sense respectively. The weak sense referred to philosophical foundations which provide readily available platforms for integrating qualitative and quantitative methods, data and designs for the purposes of problem-solving. Pragmatism was identified as belonging to this category. On the other hand, the moderate sense category comprised of philosophical foundation which articulated potent reasons for applying mixed methods in research projects. According to Shan (2022), the transformative position presented previously belongs to this category as it provides a cogent rationale for the deployment of a variety of methods (qualitative and quantitative) to accommodate various perspectives, particularly from underrepresented populations, as it relates to the phenomenon being studied. The third category, strong sense, comprises philosophical foundations which provide a robust rationalisation for the mixing of methods to achieve optimal results. An example of philosophical foundations situated in this category is dialectical pluralism (Shan, 2022).

From the foregoing and the authors' in-depth understanding of the built environment research context, the utility of pragmatism and critical realism as underlying philosophies for facilitating successful mixed methods research projects therein can be discerned. However, the prevalence of pragmatism in built environment mixed methods research happens to be the norm, compared to critical realism. Perhaps this reality can be attributed to the notion that researchers involved with mixed methods research using the pragmatic philosophical stance do not need to provide a comprehensive reason for adopting the philosophical stances besides the basic justification of selecting methods based on the nature of the research problem being studied (weak sense). Considering the implications of critical realism for engendering the attainment of the previously mentioned purposes of mixed methods research, it is surprising that studies leveraging this philosophical stance are rather scant. This can be attributed to the poor understanding of the philosophical stance among built environment researchers as the descriptions of this philosophical stance as rendered by scholars indicate its utility in carrying out either explanatory, exploratory, convergent, embedded or adaptive mixed methods studies in the built environment. Built environment research is mostly premised on the existence of an objective reality which is viewed differently by various stakeholders depending on their experiences with the phenomenon. Accordingly, the use of mixed methods to elicit these diverse perceptions stands to be supported by the critical realism philosophical stance. However, it would be difficult to establish any of these suppositions as most studies in peer-reviewed articles using mixed methods research designs are largely silent on the philosophical justification of the choice of the design. There is a need for the elucidation of the utility of critical realism and any other philosophical stances in facilitating successful mixed methods studies in the built environment using robust philosophical foundations to justify the choice of research design. This would boost the credibility of the findings emanating from such studies.

Conclusion

This chapter sets out to elucidate the significance of philosophical positioning in enabling the successful execution of mixed methods research in the built environment knowledge domain. Selecting the appropriate philosophical stance has been shown to engender the selection of the right methodology (research design) to study a particular phenomenon. This understanding has made an appreciation of the nexus between philosophical stance and the credibility of the findings from mixed methods research, as presented in this chapter, imperative.

The chapter highlights the existence of a multiplicity of philosophical stances which have been deployed in mixed methods research for different purposes. In furtherance to this, the classification of these philosophies by various authors according to the utility and strength of associated philosophical foundations for the use of mixed methods was articulated herein. Available evidence from relevant literature indicated an overt reliance on pragmatism as an underlying philosophical stance by built environment researchers involved with mixed methods research when compared to critical realism which has been acknowledged as a veritable philosophical stance for achieving the same purpose through mixed methods research.

The chapter helps to allay certain assumptions within the realm of mixed methods research in the built environment. First, the mixed methods research designs must be underpinned by a pragmatist philosophical stance. Critical realism has been shown to constitute a veritable philosophical stance, which can also underpin mixed methods research in the built environment. Lastly, methodologies relating to different philosophical stances cannot be mixed in a single study. This has been rebuffed by the existence of dialectic (pluralist), transformative and complementary strength stances respectively.

It is hoped that the contents of this chapter will deepen the discourse around the determination and selection of appropriate philosophical stances for justifying the adoption and use of mixed methods research design in the built environment.

References

Adcroft, A. & Willis, R. 2008. A snapshot of strategy research 2002–2006. *Journal of Management History*, 14(4): 313–333.

Alise, M.A. & Teddlie, C. 2010. A continuation of the paradigm wars? Prevalence rates of methodological approaches across the social/behavioural sciences. *Journal of Mixed Methods Research*, 4(2): 103–126.

Almpanis, T. 2016. Using a mixed methods research design in a study investigating the heads' of e-learning perspective towards technology enhanced learning. *The Electronic Journal of e-Learning*, 14(5): 301–311.

Ardalan, K. 2009. Globalization and culture: Four paradigmatic views. *International Journal of Social Economics*, 36(5): 513–534.

Baškarada, S. & Koronios, A. 2018. A philosophical discussion of qualitative, quantitative, and mixed methods research in social science. *Qualitative Research Journal*, 18(1): 2–21. https://doi.org/10.1108/QRJ-D-17-00042

Bhaskar, R. 1975. Forms of realism. *Philosophica*, 15(1): 99–127. https://doi.org/10.21825/philosophica.82713

Blaikie, N. 2007. *Approaches to Social Enquiry: Advancing Knowledge*. Polity Press, Cambridge.

Brennan, L., Voros, J. & Brady, E. 2011. Paradigms at play and implications for validity in social marketing research. *Journal of Social Marketing*, 1(2): 100–119.

Brierley, J.A. 2017. The role of a pragmatist paradigm when adopting mixed methods in behavioural accounting research. *International Journal of Behavioural Accounting and Finance*, 6(2): 140–154.

Brown, L. 2021. A philosophy, a methodology, and a gender identity: Bringing pragmatism and mixed methods research to transformative non-binary focused socio-phonetic research. *Toronto Working Papers in Linguistics (TWPL)*, 43(1): 2–13. https://doi.org/10.33137/twpl.v43i1.35965

Bryman, A. 2006. Paradigm peace and the implications for quality. *International Journal of Social Research Methodology*, 9(2): 111–126.

Burke, M.E. 2007. Making choices: Research paradigms and information management: Practical applications of philosophy in IM research. *Library Review*, 56(6): 476–484.

Carter, S.M. & Little, M. 2007. Justifying knowledge, justifying method, taking action: Epistemologies, methodologies, and methods in qualitative research. *Qualitative Health Research*, 17(10): 1316–1328.

Chen, H.T. 2006. A theory-driven evaluation perspective on mixed methods research. *Research in the Schools*, 13(1): 75–83.

Creswell, J.W. 2009. *Research Design: Qualitative, Quantitative and Mixed Methods Approaches*. 3rd edition. Sage, London.

Denzin, N.K. & Lincoln, Y.S. 2005. Introduction: The discipline and practice of qualitative research. In: N.K. Denzin & Y.S. Lincoln (eds). *The Sage Handbook of Qualitative Research*, pp. 1–32. 3rd edition. Sage, Thousand Oaks, CA.

Fidel, R. 2008. Are we there yet? Mixed methods research in library and information science. *Library & Information Science Research*, 30: 265–272.

Grix, J. 2004. *The Foundations of Research*. Palgrave, MacMillan, London.

Guba, E. & Lincoln, Y. 1988. Do inquiry paradigms imply inquiry methodologies? In: D. Fetterman (ed). *Qualitative Approaches to Evaluation in Education*, pp. 89–115. Praeger, New York.

Johnson, R.B. & Onwuegbuzie, A.J. 2004. Mixed methods research: A research paradigm whose time has come. *Educational Researchers*, 33(7): 14–26.

Kaushik, V. & Walsh, C.A. 2019. Pragmatism as a research paradigm and its implications for social work research. *Social Science*, 8, 255.

Kral, M.J., Links, P.S. & Bergmans, Y. 2012. Suicide studies and the need for mixed methods research. *Journal of Mixed Methods Research*, 6(3): 236–249.

Krauss, S.E. 2005. Research paradigms and meaning making: A primer. *The Qualitative Report*, 10(4): 758–770.

Mack, L. 2010. The philosophical underpinnings of educational research. *Polyglossia*, 19: 5–11.

Masadeh, M.A. 2012. Linking philosophy, methodology, and methods: Toward mixed model design in the hospitality industry. *European Journal of Social Sciences*, 28(1): 128–137.

Maxwell, J.A. & Mittapalli, K. 2010. Realism as a stance for mixed methods research. In: A. Tashakkori & C. Teddlie (eds). *SAGE Handbook of Mixed Methods in Social & Behavioral Research*, pp. 145–168. SAGE Publications, Inc., Thousand Oaks, CA.

Modell, S., 2009. In defence of triangulation: A critical realist approach to mixed methods research in management accounting. *Management Accounting Research*, 20(3), pp. 208–221.

Morgan, D.L. 2007. Paradigm lost and pragmatism regained: Methodological implications of combining qualitative and quantitative methods. *Journals of Mixed Methods Research*, 1: 48–76.

Mukumbang, F.C. 2023. Retroductive theorizing: A contribution of critical realism to mixed methods research. *Journal of Mixed Methods Research*, 17(1): 93–114.

Onwuegbuzie, A.J. & Johnson, R.B. 2006. The validity issue in mixed research, research in the schools. *Mid-South Educational Research Association*, 13(1): 48–63.

Pansiri, J. 2005. Pragmatism: A methodological approach to researching strategic alliances in tourism. *Tourism and Hospitality Planning & Development*, 2(3): 191–206.

Ruona, W.E.A. & Lynham, S.A. 2004. A philosophical framework for thought and practice in human resource development. *Human Resource Development International*, 7(2): 151–164.

Ryba, T.V., Wiltshire, G., North, J. & Ronkainen, N.J. 2022. Developing mixed methods research in sport and exercise psychology: Potential contributions of a critical realist perspective. *International Journal of Sport and Exercise Psychology*, 20(1): 147–167.

Saunders, M., Lewis, P. & Thornhill, A. 2009. *Research Methods for Business Students*. 5th edition. Pearson Education, London.

Saunders, M., Lewis, P. & Thornhill, A. 2019. *Research Methods for Business Students*. 7th edition. Pearson Education Limited, London.

Schoonenboom, J., 2019. A performative paradigm for mixed methods research. *Journal of Mixed Methods Research*, 13(3): 284–300.

Shan, Y. 2022. Philosophical foundations of mixed methods research. *Philosophy Compass*, 17(1): e12804.

Shannon-Baker, P. 2016. Making paradigms meaningful in mixed methods research. *Journal of Mixed Methods Research*, 10(4): 319–334.

Shaw, J.A., Connelly, D.M. & Zecevic, A.A. 2010. Pragmatism in practice: Mixed methods research for physiotherapy. *Physiotherapy Theory and Practice*, 26(8): 510–518.

Tronvoll, B., Brown, S.W., Gremler, D.D. & Edvardsson, B. 2011. Paradigms in service research. *Journal of Service Management*, 22(5): 560–585.

Venkatesh, V., Brown, S.A. & Sullivan, Y.W. 2016. Guidelines for conducting mixed-methods research: An extension and illustration. *Journal of the AIS*, 17(7): 435–495.

Zachariadis, M., Scott, S. & Barrett, M. 2013. Methodological implications of critical realism for mixed-methods research. *MIS Quarterly*, 37(3): 855–879.

Zou, P.X.W., Sunindijo, R.Y. & Dainty, A.R.J. 2014. A mixed methods research design for bridging the gap between research and practice in construction safety. *Safety Science*, 70: 316–326. https://doi.org/10.1016/j.ssci.2014.07.005

3 Ethical Considerations in Mixed Methods Research Design

Abid Hasan

Summary

Mixed methods research in the built environment discipline may involve both direct and indirect interactions with human participants as well as the use of data about humans. The lack of oversight or deficiencies in exercising ethical responsibilities in mixed methods research could have profound moral, professional, financial and legal implications for study participants, researchers, institutions and other stakeholders. Therefore, built environment researchers must rigorously follow ethical norms and practices in all phases of mixed methods research. However, the ethical requirements in mixed methods research can be more complicated than mono-method research. The ethical considerations that apply to different stages of a mixed methods research study are discussed in this chapter. Identifying various ethical concerns and following the best ethical research practices could help built environment researchers to navigate the ethical challenges of mixed methods research design successfully.

Introduction

Research designs guide the decisions regarding data collection, analysis, interpretation and reporting (Creswell, 2014). A mixed methods research design includes more than one research approach, which can produce converging, complementary or divergent results, increasing the rigour of the research (Heale & Forbes, 2013). Moreover, tensions between the values and processes of qualitative and quantitative paradigms could generate new insights (Lingard *et al.*, 2008). Johnson *et al.* (2007: 123) offered the following definition of mixed methods research after examining 19 definitions collected from leaders in mixed methods research.

> Mixed methods research is the type of research in which a researcher or team of researchers combines elements of qualitative and quantitative research approaches (e.g. use of qualitative and quantitative viewpoints, data collection, analysis, inference techniques) for the broad purposes of breadth and depth of understanding and corroboration.

DOI: 10.1201/9781003204046-3

Researchers use mixed methods research design in multiple ways depending on the purpose of the research, the type of qualitative and quantitative methods, and their relative arrangement and weighting in the study to provide detailed insights and comprehensive coverage of a topic of interest. Approximately 40 different types of mixed methods designs have been reported in the literature (Tashakkori & Teddlie, 2010). However, it is important to note that a mixed methods study would involve mixing methods within a single study or a series of studies investigating the same underlying phenomenon (Johnson *et al.*, 2007; Leech & Onwuegbuzie, 2009). Table 3.1 shows examples of built environment research in which mixed methods were used.

While the benefits of using a mixed methods research design in built environment research are apparent, using more than one research method can present additional ethical challenges to those in mono-method research. For instance, research participants may be asked to provide identifying information if the researcher needs to contact them later for the next phase of the study (Stadnick *et al.*, 2021). Moreover, data collection in mixed methods research could place a higher burden

Table 3.1 Examples of built environment research using mixed methods

Study	*Brief description of mixed methods research design*
Jacobsson and Linderoth (2012)	A mixed methods approach (web-based survey, participant observations and semi-structured interviews) was used to explore users' general perceptions of the impacts of information and communication technology (ICT) and analyse the implications for construction management practice.
Hallowell *et al.* (2013)	The findings regarding leading indicators of safety were cross-validated using three distinct research efforts – case studies, content analysis of reports and brainstorming sessions with an expert panel.
Abuelmaatti and Ahmed (2016)	Literature review, case studies and questionnaire surveys were used to identify factors that can enhance working collaboratively between large companies and small and medium enterprises in the construction industry.
Ijasan and Ahmed (2016)	Semi-structured interviews were used first to identify the additional factors affecting international students' housing needs apart from those discovered in the literature. Next, surveys were used to explore the needs of the students. Lastly, another round of interviews was conducted to validate the findings.
Hasan *et al.* (2021)	A literature review, focus group and questionnaire survey were conducted to identify the factors affecting the post-adoption usage of mobile ICT by construction management professionals. A literature review and focus group discussions were used to inform the questionnaire and to validate the findings.

Source: Original.

on both participants and researchers in terms of their time, effort, commitment and other resources (Stadnick *et al.*, 2021). Consequently, the use of mixed methods research design must be questioned in studies where adding a quantitative method to qualitative research or vice-versa does not advance the understanding of a given problem (Hesse-Biber, 2010).

In addition, researchers with expertise in one particular approach (e.g. quantitative- or qualitative-focused researchers) might find it challenging to implement ethical considerations in mixed methods research because they are less familiar with the other research approaches. Similarly, the ethical considerations in mixed methods research involving multi-disciplinary teams could be more complex because of differences in codes of ethics, research conduct and disciplinary cultures (Stadnick *et al.*, 2021). In addition, participant-researcher and research team relationships in mixed methods research could be far more complicated (Preissle *et al.*, 2015). Furthermore, novel and complex mixed methods research designs demand greater transparency of methodological considerations (Cain *et al.*, 2019). For instance, the approach to sampling and data validation could be very different during the quantitative and qualitative phases. As a result, the problems of representation and integration in mixed methods research could lead to specific issues of legitimation that are usually not associated with mono, or multi-method research designs (Onwuegbuzie & Johnson, 2006).

The aim of the chapter is to provide insights into ethical principles that apply to a typical mixed methods research design used in built environment research. In the chapter, ethical considerations are discussed from the perspectives of principles, research life-cycle stages and selected research methods. The main ethical principles that apply to mixed methods research are discussed first. Specific ethical considerations during the research project life-cycle are examined next. The key ethical considerations in the context of commonly used research methods in the built environment discipline are also examined. Finally, a framework for ethical research in mixed methods research design is presented in the implications section, and the key points of the chapter are summarised in the conclusion.

Ethical Considerations in Mixed Methods Research Design – A Principles-Based Approach

Ethics, derived from the Greek word *ethos* (meaning habit, character or disposition), apply to many different aspects of personal and professional settings (Aguinis & Henle, 2002). In the context of research, ethics refer to "a set of moral principles which aims to prevent the research participants from being harmed by the researcher and the research process" (Liamputtong, 2011: 25). Stewart *et al.* (2017) discussed a five-item framework (including autonomy, beneficence, non-maleficence, confidentiality and integrity) that constitutes ethically sound research in construction management involving human participants. Essentially, compliance with human ethics requires research design to be peer-reviewed and reflect the three Belmont principles of ethical research i.e. autonomy or respect for persons, beneficence and justice (Levine, 1986; Rafi & Snyder, 2015; Hasan, 2021). Therefore, it is essential to understand the following fundamental, ethical research principles and processes that apply to a mixed methods research design.

Research Integrity

Research integrity is a primary ethical concern in mixed methods research (Johnson & Onwuegbuzie, 2004). It is important to remember that "more is not always better" in research study designs, and not all research questions demand both qualitative and quantitative investigations. Therefore, researchers must demonstrate the suitability of mixed methods research for the study. Bryman (2006) examined 232 mixed methods research articles published in the Social Science discipline to identify the most common reasons for mixed methods research. The five most common reasons were enhancement (building on findings); triangulation (verifying or corroborating findings); completeness (examining the issue or topic more comprehensively); illustration (explaining findings); and sampling (identifying participants or cases in a targeted and focused manner). In built environment research, mixed methods research design has been found to improve the reliability and validity of the data and generalisation of the findings (Abowitz & Toole, 2010). However, researchers must be aware of the pitfalls of using mixed methods research design. For instance, failing to integrate quantitative and qualitative research data or reconciling contradictory mixed data can be problematic in concurrent triangulation mixed methods designs. Therefore, research integrity in mixed methods research demands clarity in planning and conducting research and a detailed and transparent depiction of the research process, including data mixing and integration details, and full disclosure of the criteria used to resolve discrepancies (Preissle *et al.*, 2015). Furthermore, researchers have an ethical obligation to report honest findings without manipulation and misinterpretation.

Beneficence

Beneficence refers to maximising benefits to participants and society while minimising possible risks, including invasion of privacy and other forms of harm – physical, psychological, social and economic. Mixed methods research might lead to an extra burden on study participants, causing stress and frustration. While sequential designs seek the involvement of participants over a long period, concurrent designs might overwhelm participants because of their intense involvement over a shorter period (Preissle *et al.*, 2015). Consequently, one of the essential ethical principles is to conduct research of which the findings are likely to benefit the discipline, broader built environment community and society.

Various disruptions, such as border closures, funding cuts and employment uncertainties during the COVID-19 pandemic, affected both built environment researchers and study participants and made them more vulnerable to harm as a result of heightened physical and mental health stressors. Therefore, the design and conduct of the research must also take contextual factors into account that might present higher than usual risks. In addition, research participants experiencing health difficulties or severe constraints on their available time could be overly burdened in mixed methods research (Creswell *et al.*, 2011). Researchers must attempt to minimise the risks to participants through a careful study design, such as developing brief instruments or intentionally modifying the research design to make it efficient, offering appropriate compensation to study participants, and clearly stating time and effort commitments in consent documents (Stadnick *et al.*, 2021).

Justice

Justice refers to the fair selection and representation of study participants to balance benefits and burdens in research (Brakewood & Poldrack, 2013). Owing to resource constraints and the lack of a publicly available database of built environment workers, researchers often rely on company websites, networking events and social media platforms to recruit participants. As a result, active users of these online platforms might become over-exposed to participation requests. In contrast, people with a limited online presence or internet reach (e.g. remote construction workers) might receive fewer opportunities to voice their opinions in mixed methods research. The over- and under-representation of different groups of participants could also affect the generalisability and scientific validity of the findings. Therefore, researchers must consider creating options and choices for less accessible target populations who might consider participating in their study. For example, the choice of online and offline participation in mixed methods research could balance the needs of different individuals and help to ensure their fair representation in research.

Informed Consent

Informed consent is the essence of human ethics. To obtain informed consent, the participants must be provided with complete information on the purpose of the study, the nature and extent of their participation, potential harms and risk management processes available, and data storage and reporting protocols (Preston, 2009; Peterson, 2013). Obtaining informed consent in mixed methods research can be challenging because of the different data collection methods and the emergent nature of the research. For instance, the questions and original protocol may change and unanticipated or distressing topics might arise during the course of the study based on initial findings in a sequential mixed methods research (Creswell *et al.*, 2011). Moreover, it could be challenging to articulate the ethical issues concerning different methods clearly in the informed consent form.

As part of the informed consent process in mixed methods research, the participants must be informed about the burden of participating in both qualitative and quantitative research if the participants are to be drawn from the same sample. Moreover, if the researcher intends to contact participants at a later time or follow up on initial results for more information, they should inform participants about the full nature of their participation (Creswell *et al.*, 2011). Participants should also be informed about the possible future use of their data in secondary data research (Hasan, 2021). Additionally, the right to withdraw from the study, and the withdrawal process, should be outlined clearly in the consent form. Furthermore, if there is any possibility of harm, potential participants must be informed and advised about the nature of harm and the help resources available (Preston, 2009). If participants are minors or from a vulnerable group (e.g. people incapacitated by disease or other conditions), then institutional rules concerning consent must be followed.

Confidentiality and Anonymity

Confidentiality assures that responses will not be publicly identified even though the researcher is aware of the identity of individual study participants. On the other hand, anonymity occurs when researchers cannot associate responses with specific participants (Preston, 2009). In mixed methods research, more data and information are generally collected from the participants and combined using different methods, which could increase their chances of being identified (Preissle *et al.*, 2015). Moreover, researchers may collect identifying information from participants to contact them for more information or a follow-up phase (Creswell *et al.*, 2011). Therefore, the processes of data collection, storage and reporting in mixed methods research must assure that the identity of the participants would not be disclosed unless failure to share information creates danger (DiCicco-Bloom & Crabtree, 2006; Peterson, 2013). Furthermore, the steps taken to maintain confidentiality and anonymity at different stages of the research must be outlined in the consent process (Holloway & Wheeler, 2010). For example, DiCicco-Bloom and Crabtree (2006) recommend that recorded data be destroyed once transcription or analysis is complete because the human voice is an identifiable form of data. Researchers should follow the institutional guidelines concerning storage, access, retention, future use and destruction of research data.

Validity

Ethics is deeply intertwined with methodological decision-making (Cain *et al.*, 2019). Researchers in the built environment must consider the distinct limits of generalisability, validity and reliability of different research methods when combining methods in a mixed methods study (Abowitz & Toole, 2010). Data validation to establish the credibility, rigour and trustworthiness of the mixed methods research design is an important ethical concern (Onwuegbuzie & Johnson, 2006; Torrance, 2012). Although it is argued that using mixed methods research design improves the validity and reliability of the resulting data and strengthens causal inferences (Abowitz & Toole, 2010), combining quantitative and qualitative methods with complementary strengths and non-overlapping weaknesses makes validity a complex issue in mixed methods research design (Onwuegbuzie & Johnson, 2006). For example, triangulation can also result in contradictions. In such cases, researchers have an ethical responsibility to explore disagreements and represent the varied perceptions or unbiased views of the phenomenon being studied (Preissle *et al.*, 2015).

Onwuegbuzie and Johnson (2006) provide a detailed account of various types of legitimation to form meta-inferences in a mixed methods research study. These types include sample integration, inside-outside, weakness minimisation, sequential, conversion, paradigmatic mixing, commensurability, multiple validities and political. However, constructing meta-inferences in mixed methods research can be problematic if differences in the study population, sampling method, sample size, generalisation claims, weaknesses and strengths of each approach, paradigm assumptions and competing dualisms are ignored while synthesising inferences from the results of the qualitative and quantitative phases (Onwuegbuzie & Johnson, 2006).

Ethical Considerations in Mixed Methods Research Design – A Life-Cycle Approach

Ethical considerations apply to all different phases of the life-cycle of a mixed methods research study, from research planning to data collection, analysis, reporting and storage. Once the choice of mixed methods research is determined, researchers should pay attention to ethical considerations in the research methods to be used in the study. The focus should be on appropriate sampling strategy, participant recruitment, development of data collection instruments and collecting data for analysis. Built environment researchers use both probability (e.g. random sampling) and non-probability techniques (e.g. convenience sampling and purposive sampling), although the use of the latter is more common in published literature (Abowitz & Toole, 2010). However, different sampling and selection methods present unique ethical challenges. For example, probability or random sampling might be theoretically possible but not feasible in built environment research because of practical considerations. Similarly, purposive sampling has the ethical challenge of assuring representation. Moreover, non-probability sampling methods do not enable the calculation of sampling error or meaningful confidence intervals (Abowitz & Toole, 2010). Consequently, implementing probability and purposive sampling techniques together within a mixed methods study could result in sampling imbalances if the researcher seeks to achieve both representativeness and data saturation (Teddlie & Yu, 2007).

Next, the data is collected after obtaining informed consent from the study participants while following the principles of beneficence and justice. After data analysis and validation, the findings are published in built environment journals, conferences or other forms of media, such as books and reports, to disseminate the results. While reporting the findings, researchers must show reflexivity and transparency about their mixed methods research design, providing a detailed account of the ethics approval process, data collection instruments, sampling procedure and bias, location and timeframe of data collection, study participants, data cleaning and analysis techniques, interpretation and ethical guidelines followed during the process (Boslaugh, 2007; Cain *et al.*, 2019). Researchers must also protect the identity of study participants using acceptable and effective data anonymisation methods in publications unless otherwise agreed upon in the consent form and other ethics approval forms.

Finally, researchers should adhere to ethical data storage practices that meet institutional and country-specific requirements. Sometimes researchers may be required to retain the original data for a certain period before destroying it, in which case, researchers must maintain the privacy and confidentiality of participants and the anonymity of stored data. If the original participants consent to the use of the data for secondary research, the data could be stored in the anonymised form beyond the institutional deadline. It is essential to pay attention to data security measures amidst the increasing number of cases of data security breaches and cyber theft. Moreover, online data collection platforms and cloud storage servers are often located in different countries and are governed by their local laws. In such cases, keeping the collected data on local, password-protected computers and storage devices can be a more ethical choice. In a few journals, the authors are encouraged to share the original data to enhance transparency. Such decisions must be guided by the conditions of the informed consent document and other ethical requirements.

Some of the ethical concerns that apply to different life-cycle stages of mixed methods research are shown in Table 3.2, compiled from various sources (DiCicco-Bloom & Crabtree, 2006; Onwuegbuzie & Johnson, 2006; Leahey, 2007; Preston, 2009; Brazeley, 2010; Peterson, 2013; Creswell, 2014; Preissle *et al.*, 2015). It is an indicative list, not an exhaustive list, to draw the readers' attention to important ethical considerations. It is recommended that readers obtain or develop a detailed research ethics checklist based on their institutional and national ethical guidelines to assist them in developing formal ethics applications and conducting ethical research.

Table 3.2 Example of ethical considerations at different stages of mixed methods research design

Mixed methods research stage	*Examples of ethical considerations*
Conceptual or research planning stage	Is qualitative or quantitative research alone insufficient to fully understand the problem? Does the research topic or questions have some scholarly or professional significance? Could participation distress study participants? Does the value of the research results justify the burden on selected participants? Can the research design be simplified to reduce the burden on participants? Has the researcher planned when and how different methods and data will be mixed? Does the research team possess the technical capacity to carry out each study component, i.e. expertise in quantitative and qualitative methods?

(*Continued*)

Table 3.2 (Continued)

Mixed methods research stage	*Examples of ethical considerations*
Participant recruitment and data collection	Are the sampling techniques compatible? Is it possible to answer the research questions with the responses from the sample size selected? Are participants coerced into consenting? Does informed consent include full disclosure about the time and effort required of study participants and how their contributions may be represented or omitted? Do the researchers need informed consent from the original participants if the data are to be used in secondary research? Are data-sharing agreements available to strengthen ethical practices? Are privacy, confidentiality and anonymity concerns respected in the recorded data?
Data analysis and reporting	Do the analysis and reporting preserve privacy and confidentiality? Does the integration of data from different data sources increase the chances of identifying study participants? Have data validity and integrity concerns been addressed during analysis when data are mixed or transformed? Is one method favoured over another in the analysis leading to bias, as the quantitative component dominates in the built environment research? Are sample sizes balanced and reasonable to support the claims made while reporting the findings? Are researchers accurate, truthful and transparent in their findings, claims and interpretations? How do the researchers present, reconcile and report the converging, partially consistent or contradictory findings or different interpretations of the data? Do the publications include full disclosure of research design, methods and results? Are the participants and their views represented respectfully and fairly without placing them at risk? Do the researchers take the steps necessary to avoid misinterpretations of the results?
Data storage	Does the method of data storage preserve privacy and confidentiality? Are robust data security measures used? Are the stored data in an anonymised form? Can people outside the research team access stored data? Did the researchers obtain consent for future use of data in secondary research?

Source: Original.

Ethical Considerations in Mixed Methods Research Design – A Methodological Approach

In mixed methods research design, many ethical considerations depend on the specific methods of data collection used in the research. Since most aspects of the built environment involve people, social interactions and social processes (Abowitz & Toole, 2010), researchers often use combinations of different research methods of Social Sciences. In view of the vast variety of research methods, the ethical concerns regarding focus groups, interviews and surveys only are discussed briefly in the following sections. The author has discussed the ethical considerations in secondary data research elsewhere (Hasan, 2021).

Focus Groups

Focus groups are often used in mixed methods research studies in the built environment to inform the development of survey or interview questions and to validate the data collected with the use of other instruments. The main ethical concern is that the researcher cannot ensure that all discussions in the focus group will remain confidential (Smithson, 2008). Regardless of whether the group is composed of strange or familiar faces, individuals might feel uncomfortable discussing their concerns in a group context (Smithson, 2008). Additionally, group members might voice opinions that are distressing to other participants (Kitzinger & Barbour, 1999). When discussing sensitive topics, group members might not respond sensitively to the personal disclosures of others. As a result, participants might reveal more personal and intimate details about their lives than they should (Pini, 2002) and later become uncomfortable about their over-disclosure (Smith, 1995). In such cases, researchers should warn participants about the possibility that such disclosure might happen (Smith, 1995).

Interviews

Considering one-to-one interaction, an intrusion into interviewees' lives and the sensitivity of the interviewing process, a high standard of ethical considerations should be maintained at all stages of the interview process (Doody & Noonan, 2013; Alshenqeeti, 2014). DiCicco-Bloom and Crabtree (2006) recommend that four ethical issues should be considered in the interview process: (1) reducing the risk of unanticipated harm, (2) protecting the interviewee's information, (3) informing interviewees about the nature of the study and (4) reducing the risk of exploitation. The interviewer should be able to identify non-verbal behaviour and respond to situations where a participant is feeling uncomfortable in responding to a question by moving away from the topic, rephrasing the question or, in some cases, pausing or ending the interview (Doody & Noonan, 2013). Moreover, because of the small sample sizes used in qualitative research, the risk that individuals with distinctive characteristics can be identified from the description of interviewees or verbatim quotes from the interview is significant. Therefore, more attention should be paid to confidentiality and anonymity requirements.

Online Surveys

As online or web-based surveys have gained popularity in built environment research, various settings for the web-based survey must be selected carefully when designing a questionnaire to ensure the integrity and confidentiality of both data and the respondents. For example, the respondents' privacy and anonymity could be maintained by using the built-in privacy options of survey platforms that do not allow identifying information, such as email or Internet Protocol addresses to be recorded. In addition, the completed questionnaires could be numbered independently of any tracking scheme to ensure the anonymity of respondents. While a questionnaire survey could be conducted anonymously in a mono-method quantitative study, participants may be asked to reveal identifying information for a qualitative follow-up phase in a sequential mixed methods study (Creswell *et al.*, 2011). In such cases, the privacy and identities of study participants must be respected by implementing various safeguards. Moreover, questions should be designed in such a way that they are not leading or invasive, contain implicit assumptions or invite respondents to breach confidentiality. In online surveys, it is recommended that an opening statement or a cover letter is used to clarify that consent will be considered to be implicit if participants respond to the survey (Thwaites-Bee & Murdoch-Eaton, 2016). In addition, the online recruiting strategy should follow online etiquette, such as not spamming or contacting potential study participants excessively.

Implications

The emphasis on ethical conduct in built environment research has increased considerably in recent years. In many countries, researchers in the built environment must comply with national and institutional research ethics, regulations and codes of conduct for ethical research. They are required to submit a formal ethics application to an ethics committee or an internal review board before starting the data collection. They are also required to submit research progress reports at regular intervals to update the committee on the status of the research, compliance with ethical requirements and ethical concerns (if any). In addition, many built environment journals now ask for details of ethics application approval and declarations concerning ethical research conduct when submitting a manuscript.

The use of mixed methods research design in the built environment discipline offers several opportunities for researchers to investigate complex and multi-layered research problems in more depth, using both quantitative and qualitative research methods. However, as previously discussed in the chapter, planning for mixed methods research, collecting and integrating multiple data sources and reporting the findings require careful ethical considerations to reduce the likelihood of ethical issues and harm to study participants and researchers. Consequently, novice researchers and graduate students could be at higher risk of breaching their ethical responsibilities and academic integrity when using mixed methods research design. Therefore, built environment researchers undertaking mixed methods research must carefully understand and implement ethical principles and guidelines applicable to different stages of the investigation. Insights into ethical considerations in mixed methods research design were offered in the chapter from three perspectives – principal-based and life-cycle and methodological approaches – as shown in Figure 3.1.

Conclusions

It is essential to consider the ethical research norms and procedures as an integral part of mixed methods research design. Ethical principles that apply to mixed methods research design in the built environment discipline were discussed in

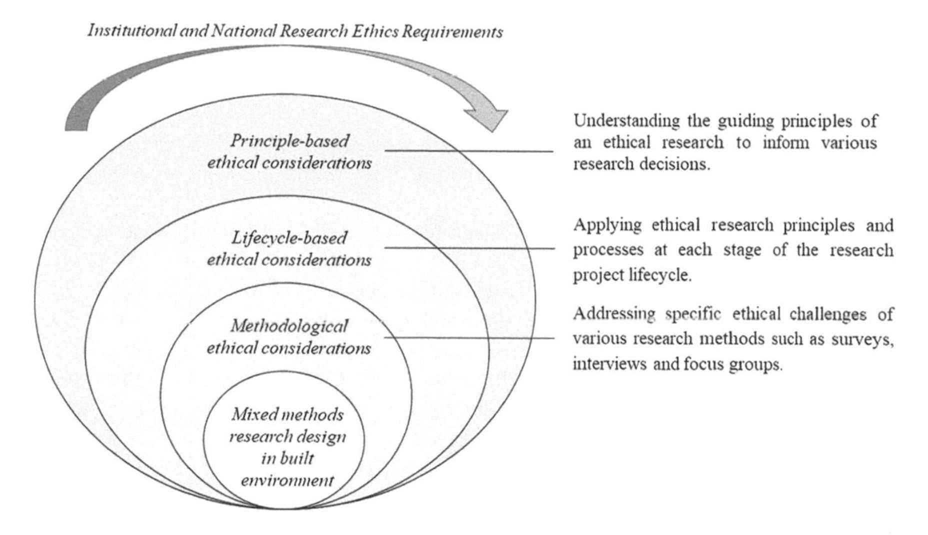

Figure 3.1 A framework of ethical consideration in mixed methods research design

the chapter. Readers must note that summarising all the ethical requirements of a mixed methods research design was beyond the scope of this chapter. Moreover, it was not the intention of the chapter to advise on the specific institutional and national ethical frameworks and guidelines to which built environment researchers must adhere while conducting mixed methods research. Therefore, readers must consult their institutional and national ethical research framework and guidelines and discuss their ethical concerns with their colleagues, supervisors and other more experienced researchers. Moreover, books on specific research methods should be consulted to understand fully the ethical considerations that apply to different data collection instruments. Nonetheless, the information presented in the chapter is expected to broaden the understanding of ethical considerations that apply to mixed methods research in the built environment discipline. Researchers in the built environment must consider specific ethical concerns regarding beneficence and justice, sampling, confidentiality and anonymity, informed consent, validity and overall research integrity during the research project life-cycle and across various research methods used in their mixed methods research.

References

Abowitz, D.A. & Toole, T.M. 2010. Mixed method research: Fundamental issues of design, validity, and reliability in construction research. *Journal of Construction Engineering and Management*, 136(1): 108–116.

Abuelmaatti, A. & Ahmed, V. 2016. Enabling building-information-modelling-capable small and medium enterprises. In: V. Ahmed, A. Opoku & Z. Aziz (eds). *Research Methodology in the Built Environment*, pp. 167–188. Routledge, London.

Aguinis, H. & Henle, C. 2002. Ethics in research. In: S.G. Rogelberg (ed). *Handbook of Research Methods in Industrial and Organizational Psychology*, pp. 34–56. Blackwell Publishing, Malden, MA.

Alshenqeeti, H. 2014. Interviewing as a data collection method: A critical review. *English Linguistics Research*, 3(1): 39–45.

Boslaugh, S. 2007. *Secondary Analysis for Public Health: A Practical Guide*. Cambridge University Press, Cambridge, NY.

Brakewood, B. & Poldrack, R. 2013. The ethics of secondary data analysis: Considering the application of Belmont principles to the sharing of neuroimaging data. *Neuroimage*, 82: 671–676.

Brazeley, P. 2010. Computer-assisted integration of mixed-methods data sources and analyses. In: A. Tashakkori & C. Teddlie (eds). *Mixed Methods in Social and Behavioral Research*. 2nd edition, pp. 431–468. Sage, Thousand Oaks, CA.

Bryman, A. 2006. Integrating quantitative and qualitative research: How is it done? *Qualitative Research*, 6(1): 97–113.

Cain, L.K., MacDonald, A.L., Coker, J.M., Velasco, J.C. & West, G.D. 2019. Ethics and reflexivity in mixed methods research: An examination of current practices and a call for further discussion. *International Journal of Multiple Research Approaches*, 11(2): 144–155.

Creswell, J.W. 2014. *Research Design: Qualitative, Quantitative, and Mixed Methods Approaches*. 4th edition. Sage Publications, Thousand Oaks, CA.

Creswell, J.W., Klassen, A.C., Plano Clark, V.L. & Smith, K.C. 2011. Best practices for mixed methods research in the health sciences. *Office of Behavioral and Social Sciences Research, National Institutes of Health*. [Online]. Available at: https://obssr.od.nih.gov/sites/obssr/files/Best_Practices_for_Mixed_Methods_Research.pdf

DiCicco-Bloom, B. & Crabtree, B.F. 2006. The qualitative research interview. *Medical Education*, 40(4): 314–321.

Doody, O. & Noonan, M. 2013. Preparing and conducting interviews to collect data. *Nurse Researcher*, 20(5): 28–32.

Hallowell, M.R., Hinze, J.W., Baud, K.C. & Wehle, A. 2013. Proactive construction safety control: Measuring, monitoring, and responding to safety leading indicators. *Journal of Construction Engineering and Management*, 139(10): 04013010.

Hasan, A. 2021. Ethical considerations in the use of secondary data for built environment research. In: E. Manu & J. Akotia (eds). *Secondary Research Methods in the Built Environment*, pp. 26–39. 1st edition. Routledge, London.

Hasan, A., Ahn, S., Rameezdeen, R. & Baroudi, B. 2021. Investigation into post-adoption usage of mobile ICTs in Australian construction projects. *Engineering, Construction and Architectural Management*, 28(1): 351–371.

Heale, R. & Forbes, D. 2013. Understanding triangulation in research. *Evidence-Based Nursing*, 16(4): 98–98.

Hesse-Biber, S. 2010. Qualitative approaches to mixed methods practice. *Qualitative Inquiry*, 16, 455–468.

Holloway, I. & Wheeler, S. 2010. *Qualitative Research in Nursing and Healthcare*. 3rd edition. Wiley-Blackwell, Oxford.

Ijasan, K. & Ahmed, V. 2016. Studentification and the housing needs of international students in Johannesburg: An embedded mixed methods approach. In: V. Ahmed, A. Opoku & Z. Aziz (eds). *Research Methodology in the Built Environment*, pp. 149–166. Routledge, London.

Jacobsson, M. & Linderoth, H.C. 2012. User perceptions of ICT impacts in Swedish construction companies: 'It's fine, just as it is'. *Construction Management and Economics*, 30(5): 339–357.

Johnson, R.B. & Onwuegbuzie, A.J. 2004. Mixed methods research: A research paradigm whose time has come. *Educational Researcher*, 33(7): 14–26.

Johnson, R.B., Onwuegbuzie, A.J. & Turner, L.A. 2007. Toward a definition of mixed methods research. *Journal of Mixed Methods Research*, 1(2): 112–133.

Kitzinger, J. & Barbour, R.S. 1999. Introduction: The challenge and promise of focus groups. In: R.S. Barbour & J. Kitzinger (eds). *Developing Focus Group Research*, pp. 1–20. Sage Publications, London.

Leahey, E. 2007. Convergence and confidentiality? Limits to the implementation of mixed methodology. *Social Science Research*, 36(1): 149–158.

Leech, N.L. & Onwuegbuzie, A.J. 2009. A typology of mixed methods research designs. *Quality & Quantity*, 43(2): 265–275.

Levine, R. 1986. *Ethics and Regulation of Clinical Research.* 2nd edition. Urban and Schwarzenberg, Baltimore, MD.

Liamputtong, P. 2011. *Focus Group Methodology: Principles and Practice*. Sage, London.

Lingard, L., Albert, M. & Levinson, W. 2008. Grounded theory, mixed methods, and action research. *British Medical Journal*, 337: a567.

Onwuegbuzie, A.J. & Johnson, R.B. 2006. The validity issue in mixed research. *Research in the Schools*, 13(1): 48–63.

Peterson, R. 2013. *Constructing Effective Questionnaires*. Sage, Thousand Oaks, CA.

Pini, B. 2002. Focus groups, feminist research and farm women: Opportunities for empowerment in rural social research. *Journal of Rural Studies*, 18(3): 339–351.

Preissle, J., Glover-Kudon, R., Rohan, E.A., Boehm, J.E. & DeGroff, A. 2015. Putting ethics on the mixed methods map. In: S.N. Hesse-Biber & R.B. Johnson (eds). *The Oxford Handbook of Multimethod and Mixed Methods Research Inquiry*, pp. 144–162. Oxford University Press, Oxford, UK.

Preston, V. 2009. Questionnaire survey. In: R. Kitchin & N. Thrift (eds). *International Encyclopedia of Human Geography*, pp. 46–52. Elsevier, Oxford.

Rafi, N. & Snyder, B. 2015. Ethics in research: How to collect data ethically. In: M.P. Wilson, K.Z. Guluma & S.R. Hayden (eds). *Doing Research in Emergency and Acute Care: Making Order Out of Chaos*, pp. 45–51. John Wiley & Sons Ltd., Chichester.

Smith, M.W. 1995. Ethics in focus groups: A few concerns. *Qualitative Health Research*, 5(4): 478–486.

Smithson, J. 2008. Focus groups. In: P. Alasuutari, L. Bickman & J. Brannen (eds). *The SAGE Handbook of Social Research Methods*, pp. 357–370. Sage Publications, Thousand Oaks, CA.

Stadnick, N.A., Poth, C.N., Guetterman, T.C. & Gallo, J.J. 2021. Advancing discussion of ethics in mixed methods health services research. *BMC Health Services Research*, 21(1): 1–9.

Stewart, I., Fenn, P. & Aminian, E. (2017). Human research ethics – Is construction management research concerned? *Construction Management and Economics*, 35(11–12): 665–675.

Tashakkori, A. & Teddlie, C. 2010. *SAGE Handbook of Mixed Methods in Social & Behavioral Research.* 2nd edition. SAGE Publications Inc., Thousand Oaks, CA.

Teddlie, C. & Yu, F. 2007. Mixed methods sampling: A typology with examples. *Journal of Mixed Methods Research*, 1(1): 77–100.

Thwaites-Bee, D. & Murdoch-Eaton, D. 2016. Questionnaire design: The good, the bad and the pitfalls. *Archives of Disease in Childhood, Education and Practice Edition*, 101(4): 210–212.

Torrance, H. 2012. Triangulation, respondent validation, and democratic participation in mixed methods research. *Journal of Mixed Methods Research*, 6, 111–123.

4 A Review of Mixed Methods Research Design in Construction Education

Saeed Rokooei

Summary

Research projects in construction education are dependent upon human-based systems, exploring ideas, theories, and variable associations among human actions, behaviours, or perceptions. This necessitates the effective application of social science research methods. Typically, the complexity and inter-dependencies among factors or subjects in social science research relating to construction education demand a combination of different data collection methods and research approaches to obtain reliable outcomes. In this chapter, literature about construction education published in a major construction journal over the past decade is reviewed with the objective of exploring the extent to which the knowledge generated through research considers the situational nature of educational methods in the three areas of archival studies, empirical research, and simulation models. In this chapter, deficiencies, and limitations in these studies, corresponding to features of mixed methods approaches in the reviewed papers, are mapped. The analysis reveals pitfalls that were overlooked and, therefore, have compromised the validity and reliability of these studies. Such observation provides a methodological structure, strategies, and guidelines for developing experimental frameworks, data collection methods, and modelling mixed methods approaches for research in construction education. This chapter contributes to the body of knowledge by articulating a framework for the design of different types of mixed methods approaches in construction education studies.

Introduction

Converting the knowledge and experience of construction researchers into educational facts, hypotheses, and constructs remains a challenging undertaking. Construction education covers subjects that are addressed in scholarly endeavours in the context of construction to explore methods, tools, models, and frameworks that can improve the quality of the learning process. Academic research activities within construction schools are seldom sufficient for educational purposes and, thus, there is a vital need for industry-wide data. These data are obtained or extracted from industry sources, including construction professionals, organisations,

DOI: 10.1201/9781003204046-4

equipment, materials, or methods. The data-gathering process is often undertaken through industry advisory boards and professional organisations. However, many factors affect the adequacy and quality of data collected throughout the process. The procedure to design, plan, execute, monitor, maintain, and retrieve data is a constraint for many studies in construction. Therefore, it is crucial to examine the research tools and methods to ensure that the mechanisms of data management in construction studies are appropriately and effectively employed. The management of research data, which consists of all steps mentioned above, is significantly correlated with the research methodology. The mixed methods approach is generally under-utilised despite the alignment between the characteristics of mixed methods and the nature of construction research subjects. Thus, there is a need to evaluate the extent to which research methodologies are appropriately employed in construction studies. To address this need, the use of mixed methods research design in research about construction education is reviewed in this chapter. A comparative assessment of publications in construction education indicated that the *Journal of Civil Engineering Education (JCEE)* is a prominent platform for publishing scholarly papers in construction education. The comprehensive review of papers published from 2011 to 2021 in JCEE showed deficiencies and shortcomings in some research papers. Also, the analysis of bibliographic information revealed various aspects of authorship in construction education manuscripts. In this chapter, the fitness of mixed methods in construction education is emphasised, while providing various guidelines and suggestions in the design of mixed methods research.

Background

Construction Education Research

Lack of structured data, subjectivity, lack of proper documentation mechanism, and inconsistent protocols are among the factors that affect the processing of data. In addition to these factors, the long duration of projects, multi-agent workflows, site specifications, and high turnover of construction personnel exacerbate the quality of data modelling in research about construction education (El-Diraby & O'Connor, 2004). Since many research projects in construction are related to human actions or behaviours, a social science research approach is necessary, in which problem definition, identification of factors, relationship specification, and determination of impact are not linear and simple. In this situation, using multiple or mixed methods can attenuate the shortcomings of qualitative or quantitative methods in construction research (Abowitz & Toole, 2010). In a review of 43 papers published in the *Journal of Construction Engineering and Management*, Abowitz and Toole (2010) showed that despite papers being focused on social science constructs, multiple methods were not used in any of the papers. Zheng *et al.* (2019) conducted a bibliometric review of research about construction education during the past three decades and showed that there had been a trend towards the utilisation of technologies in education, which indicated a shift towards contextual factors and topics focused on impact. Zheng *et al.* (2019) also concluded that the most construction education research had been published in ten journals. Among those journals, the *Journal of Civil Engineering Education*, formerly known as the *Journal of Professional Issues in Engineering Education and Practice (JPIEEP)*, had received the most citations and had the highest hub score (0.6265) regarding its publication of articles about construction education. In another research review conducted by Zou *et al.* (2014) on safety in construction, it was shown that quantitative methods had the highest percentage (43.2%), followed by qualitative methods (23.9%), whereas only 9.1% of publications used a mixed methods design. The high percentage of publications in which a quantitative method was used, indicated that construction researchers largely adopted an objectivist philosophical standpoint according to which safety-related issues are framed as products, not iterative processes. This statement refers to the classification provided by Creswell (2009) in which different types of methodologies were linked to various applications (Tashakkori & Teddlie, 2010). Based on this classification, mainly quantitative methods are used in post-positivist research in which the main goal is to create knowledge discovery and verification through direct observations or measurements of surrounding factors and environments. On the other hand, qualitative methods are employed in constructivist research to interpret the meaning of phenomena in studies, while this process is dependent on time and context. In addition, both quantitative and qualitative methods are used in advocacy/participatory research based on a pragmatic paradigm (Naaranoja *et al.*, 2014).

Research Methodologies in Construction

Panas and Pantouvakis (2010) reviewed 89 papers, mainly from five prominent construction journals from 1999 to 2009, to identify different research methodologies used to investigate construction productivity in construction journals. Panas and Pantouvakis (2010) developed a model of research taxonomy in construction productivity, representing three levels of classification, categories, and method focus. At the first level, classification was divided into three types: qualitative, quantitative, and mixed methods. The categories recognised at the second level were archival study, empirical research, and simulation proposals, and the method focus included experimental framework, data collection, and modelling techniques. The analysis by Panas and Pantouvakis (2010) showed that 60.7% of papers had used quantitative methods, while the share of qualitative and mixed methods approaches was 10.1% and 29.2%, respectively. Holt and Goulding (2017) reviewed the triad in terms of ontological/epistemological viewpoints and paradigmatic approaches to research in construction management and concluded that ontology and epistemology are most relevant to methodological decisions made in construction management research. Flood and Scott (2015) reviewed stages in structuring a research design and methodological approaches suited to topics in construction and identified four types of validity and associated issues, including construct validity, statistical conclusiveness validity, internal validity, and external validity. Based on these types, a research project in construction should provide indicators to capture the expected relationships among the concepts being researched. It is also necessary to address the relationships between hypothesised independent and dependent variables from a statistical point of view and show whether the difference is significant. The research should also demonstrate causal or plausible links between hypothesised variables. It is necessary to show whether relationships found within the sample's subjects are generalisable to the population. Liu *et al.* (2014) presented a comprehensive and critical review of the validation methodologies used in research about construction engineering and management, in which they presented eight categories of validation methodologies, including experimental studies, observational studies, empirical studies, case studies, survey, functional demonstration, and archival data analysis. The experimental studies include sets of experimental data, statistical analysis, and grouping. Observational studies include the use of audio-visual technologies (e.g., cameras, videos, microphones) and pre-set frameworks instead of randomisation. Empirical studies are mainly designed based on model development and evaluation as well as statistical analysis. As the name implies, case studies are focused intensively on the analysis of a particular matter or facet of a specific topic. Surveys make use of sampling and data collection through different instruments and methods. In the functional demonstration, input, assumptions, and output are validated by modelling, simulation, algorithms, machines, or programmes. Finally, archival data analysis is generally dependent on historical data that is published in certain intervals.

Research Methodology

The main goal of this study was to explore research methodologies used in construction education with a view to understanding the extent to which mixed methods research designs have been used in these methodologies. A retrospective content analysis of papers published in the *JCEE* was employed in the study. *JCEE*, among all construction-related journals, receives the most citations, has the highest hub score (0.6265), and is ranked second in the bibliographic records (Zheng *et al.*, 2019) and, therefore, is a suitable publication to investigate over a period of time. All papers that were published between 2011 and 2021 were documented, and only technical papers were selected for further analysis. The reason for the exclusion of case studies and technical notes was to ensure that collected samples generally followed a standard research design. The sample size included 254 papers. Throughout this process, the general approach of systematic review evaluation provided by PRISMA was used. Through the review of papers, possible direct statements about the research methodology used were scrutinised to confirm the type of methodology employed. In addition, three sets of keywords associated with quantitative, qualitative, and mixed methods research methods were identified through the literature review (Bloor & Wood, 2006; Tashakkori & Teddlie, 2010). These sets included exact or variations of key terms, including quantitative, survey, questionnaire, correlation, statistical, qualitative, case study, interview, indexing, observation, mixed method, triangulation design, embedded design, explanatory design, and exploratory design. The occurrence of each term in each paper was separately determined and recorded. The frequency of keywords used in papers facilitated the determination of the type of methodology when the methodology was not explained explicitly. Other information regarding the authorship of the papers was extracted and stored in the data model, such as the number of authors, themes, student authorship, and tools. All data were recorded and coded for statistical analysis.

Results

The set of samples included 254 technical papers that were published from 2011 to 2021 in the *JCEE*. While the range of papers' subjects and themes was broad, education-related topics were the most frequent keywords. Table 4.1 shows the first, second, and third most frequent terms for keywords; however, there were no keywords for any methodology method or approach. The mixed methods term appeared once, and quantitative was found twice as the keyword, while there was no mention of the qualitative method in the keywords.

The bibliographic analysis of the papers showed that three authors had the highest percentage among the total number of authors. Table 4.2 shows the percentage of the number of authors per paper.

Also, in 63% of papers, no students were among the authors, while the percentages for one student, two students, and three students as co-authors were 30%, 5%, and 2%, respectively. In addition, in 17% of papers, a student was the first author.

In the next step, a comprehensive thematic analysis was conducted to determine the frequency of research types in the published papers. Research type identification was performed in a three-stage, filtering process. The first filter was the paper self-identification. Any direct statement about the type of the research was considered the research type identifier. Statements such as "a quantitative research method was employed to identify challenges, potentials, importance, and gaps in

Table 4.1 Most frequent keywords

Frequency rank	*Keyword 1*	*Keyword 2*	*Keyword 3*	*Keyword 4*	*Keyword 5*
First	Engineering education	Engineering education	Engineering education	Engineering education	Professional development
Second	Sustainability	Curricula	Curricula	Social factors	Sustainable development
Third	Construction	Project management	Education	Sustainable development	Engineering profession

Source: Original.

Table 4.2 Percentage of the number of authors of papers

Number of authors	*1*	*2*	*3*	*4*	*5 or more*
Percentage	11	25	29	17	18

Source: Original.

students' knowledge and quantify the comparisons" (Rokooei *et al.*, 2022), "the authors conducted a qualitative inquiry that considered a traditional school and work cultures within the US university system" (Groen *et al.*, 2019), and "through a robust mixed methods approach with positive quantitative results, this study aimed to serve as a trusted and transferrable source for university educators" (Clark *et al.*, 2021) were examples of such direct statements. In the next step, the frequency of keyword assessment was applied to distinguish different types of research methods. For this purpose, common keywords for each research type were extracted from the literature. The keywords for quantitative methods included "quantitative", "survey", "questionnaire", "correlation" and "statistical". Qualitative methods keywords consisted of "qualitative", "case study", "interview", "indexing", and "observation". Also, "mixed method", "triangulation design", "embedded design", "explanatory design" and "exploratory design" were keywords for the mixed methods approach. The third evaluation stage was conducted by reviewing the methodology and analysis sections of the papers. Depending on the structure designed for the study, data collection, and data analysis, the type of the research methodology was specified. As a result, 138, 32, and 51 papers were classified into quantitative, qualitative, and mixed methods groups, respectively. Also, 33 papers were excluded from this classification because of the nature of the content of the manuscript. Table 4.3 shows the percentage of each category.

Also, the frequency of keywords in each category was analysed, and the average occurrence of each keyword in that category was calculated. As shown in Table 4.4, mixed method keywords had a considerably lower level of occurrence in their corresponding groups.

Table 4.3 Percentage of the research type

Research type	*Quantitative*	*Qualitative*	*Mixed*	*Others*
Percentage	54	13	20	13

Source: Original.

Table 4.4 Average of keyword frequency in each group

Keyword	Quantitative	Survey	Questionnaire	Correlation	Statistical	Qualitative	Case Study	Interview	Indexing	Observation	Mixed-methods	Triangulation design	Embedded design	Explanatory design	Exploratory design
Average occurrence	1.01	16.23	3.83	2.5	4.7	4.25	2.34	19.88	0	2.34	1.16	0	0	0	0

Source: Original.

Also, the most frequent keywords referring to statistical tools or approaches were Descriptive Analysis, Comparison, t-test, Exploratory Factor Analysis, Chi-square, and Anova.

Discussion

Qualitative and quantitative research methods have advantages and disadvantages, which make either type appropriate for certain situations. Qualitative research can be used to obtain in-depth, detailed feedback and comments by encouraging discussion. However, the drawbacks of qualitative research include the small sample size, finding generalisations, and lack of anonymity. On the other hand, in quantitative research, data collection is relatively rapid and easy, samples can be selected randomly, and reliable and repeatable information can be obtained. However, follow-ups and detailed feedback are not an option. In addition, the potential for conducting the research in an unnatural environment exists in quantitative research. The review of the collection of papers published in the *JCEE* revealed that the mixed methods approach was still underutilised. While the nature of construction education studies is suitable for mixed methods research, many studies fall short of including research approaches that incorporate qualitative and quantitative aspects. Several subjects in research design are highlighted through a comprehensive review of published papers. The following items are major subjects that need the careful attention of construction scholars:

- Research question

  Defining a clear, compelling, and direct research question is critical in research design. Various sources have defined different characteristics for research questions (Creswell, 2009; Krosnick & Presser, 2009; Pandey & Pandey, 2015). Research questions should be structured. The researcher must define the scope of research meticulously and specify what areas are within the scope and what areas fall beyond that. Also, the research question should be specific to explore a certain aspect of the topic. In addition, the question subject should be explorable and relevant. The following questions are taken from the reviewed papers:

  1 Do faculty believe that civil engineering undergraduate students in their programme learn about ESI [ethics and/or the societal impacts] by participating in co-curricular activities? (Bielefeldt *et al.*, 2020: 2)
  2 Do board game design and gameplay in the classroom achieve higher levels of perceived cognitive/learning engineering-related and course-specific concepts, achieve higher levels of student engagement, create a sense of community, improve students' professional skills, increase student retention in STEM, and yield easy transferability for university-level educators? (Clark *et al.*, 2021: 1)

  The first question is relatively concise and direct, and the target population (i.e. civil engineering faculty) is clear. Also, the question is explorable and relevant as it addresses an essential aspect of the civil engineering career.

One potential issue that might emerge through the examination is the scope of the topic, which is ethics and the societal impacts. The researcher needs to ensure that there is a homogeneous perception of the subject. One way to increase the likelihood of certainty is to provide short statements and explanations about the topic. A similar discussion is somewhat applicable to the other keyword (co-curricular) as well.

The second question is longer, with multiple sections. Generally, long questions are prone to uncertain terms. In addition, a potential issue might arise when analysing the data. Such multi-segment questions can either be broken down into independent segments or considered a whole and be responded to collectively. In either case, it entails some potential issues. Analysing data and responding to several segments might deviate the flow from the main theme or finding of the study, as well as lengthen the analysis. Considering various segments under a larger topic or construct makes drawing a uniform conclusion difficult. Independent segments might be evaluated differently and, thus, combining these evaluations to present a consistent trend might not be feasible. In general, as the research goal moves from descriptive to relational and then causal, these concerns become more prominent.

- Construct definition

  Defining a construct lays a foundation for further activities in research. A construct is an abstract idea, underlying theme, or subject matter that researchers aim to explore. While constructs range from simple to complex, they influence the methodology used. Complex constructs convey various dimensions and features. These features together represent the construct by their commonality as a whole. Constructs can express ideas, individuals, structures, events, or objects. It should be noted that constructs are assessed by variables. Also, variables are measured through questions in surveys or interviews. Therefore, it is vital to select an appropriate method and analysis strategy.

- Research method

  While the selection of the research method depends on many factors, the mixed methods approach is generally well suited to construction education research. The inter-twined technical aspects and inter-personal skills in construction education make mixed methods strategies a robust approach. The triangulation of data collection enhances the credibility and validity of research findings. As discussed in the previous sections, defining variables to measure and, ultimately, form the construct needs a strong background in practice and theory. Integrating qualitative and quantitative methods can ensure the efficacy of construct and variable definition. The following excerpt from a paper authored by Alaka *et al.* (2017: 1) shows how a combination of qualitative and quantitative methods can facilitate the analysis to answer the research questions:

  This study thus set out to uncover insolvency criteria of S&M [small or micro] CEFs [civil engineering firms] and the underlying factors using mixed methods. Using convenience sampling, the storytelling method was used to execute interviews with 16 respondents from insolvent firms. Narrative and thematic analyses were used to extract 17 criteria under two groups. Criteria were used to formulate a questionnaire, of which 81 completed copies were received and analysed using Cronbach's alpha coefficient and relevance index score for reliability and ranking, respectively. The five most relevant criteria were economic recession, immigration, too many new firms springing up, collecting receivables, and burden of sustainable construction. The four underlying factors established through factor analysis were market forces, competence-based management, operations efficiency and other management issues, and information management.

- Data collection

  Data collection is a systematic approach to collecting data accurately and effectively from various sources and providing sufficient materials for data analysis. The data are obtained from various sources, including surveys, interviews, observations, focus groups, reports, records, and extracted meta-data. Each source used has advantages and challenges that affect further steps. In construction education research, a few common sources form most of the data generation, including data about student performance or perception, industry professionals' opinions, and data about device/system performance. Construction scholars often use students' feedback and perceptions in the form of pretest/posttests. In this situation, controlling other visible or latent factors becomes critical. For example, Cho *et al.* (2013) explored the effects of guided inquiry module instruction on undergraduate construction engineering students and employed different tools, including a pretest and posttest. The following excerpt shows extracts from the study:

  To determine whether students in the two sections were similar based on previous knowledge of topics covered in the class, pretest quiz scores of the two groups were compared (a pretest quiz was given at the onset of each module topic). For all course modules, Section B had significantly higher pretest scores at a level of $p < 0.001$, with effect sizes ranging from −1.08 (Aggregates) to −3.00 (Wood).

  While the original intention of the researchers was to compare the gain scores of two sections, it was determined that only posttest scores could be compared based on this existing difference between the groups.

  Instructor and treatment effects were not able to be simultaneously investigated because of the study design.

  The pretest quizzes were designed to test basic understanding of the concepts of the block of instruction, but the posttest quizzes were more rigorous in their treatment of the relevant topics.

  The limitation of the data collection process is implied in the statements above. With two sections (Section A and B), two instruction methods (traditional and guided inquiry module), and two instructors, a setting for pretest and posttest could include identical tests at the beginning and end of each module.

Any difference could indicate the effectiveness of instruction methods, possibly at two different levels. Another setting could measure student performance at any point (possibly at the end of each module) through a cross-section evaluation. This could show whether guided inquiry module instruction, as intended in the study, was more effective than the traditional method. However, within any setting, considering the properties of data collection tools and managing independent, dependent, and confounding variables are crucial.

- Data analysis

  Data analysis is at the core of research, in which statistical or logical techniques are systematically applied to determine patterns, relationships, or trends among gathered data. The form and tools utilised in the data analysis are determined by the research methodology used. In qualitative research, various techniques and tools are used to analyse the data. Content analysis, narrative analysis, discourse analysis, thematic analysis, and grounded theory are examples of such techniques. On the other hand, in quantitative research, two main categories of methods for data analysis are used. The first category is a descriptive analysis which includes measures of frequency, measures of central tendency, measures of variation, and measures of position. The second category is inferential statistics in which many statistical tools are used such as correlation, cross-tabulation, regression analysis, frequency tables, and analysis of variance. One important consideration in using any of the data analysis tools or techniques is to check their requirements. Statistical tools typically require a set of conditions as the first step and, therefore, researchers should ensure that such requirements are met. While collecting data, the normality of data should be investigated. For example, when a study compares two groups, a normal data set allows parametric tests such as t-tests. Similarly, non-parametric tests such as the Wilcoxon signed-rank test can be used for non-normal data. One of the most common flaws in construction education studies is to overlook such requirements.

- Validity and reliability

  Validity and reliability are two key terms (and concepts) in effective research design and implementation. Validity is the extent to which the scores from a measure accurately represent the variable. Reliability refers to the consistency of a measure or procedure throughout the research process. Construction education research suffers in both areas as, sometimes, these two concepts are neglected. As discussed before, variables that accurately portray concepts, subjects, or measures in construction education need meticulous definitions of content and scope. Also, educational settings impose limitations on the research design. Thus, ensuring validity and reliability in construction education research might be overlooked. Nonetheless, validity and reliability must be addressed in construction research. Reliability can be examined over time (e.g. test-retest reliability), across subjects (e.g. internal consistency), and across different researchers (e.g. inter-rater reliability). The validity can be investigated in several distinct ways, including face validity, content validity, and criterion validity. However, the evaluation of validity and reliability is an ongoing process. In the reviewed set of papers, even with counting out-of-context instances, validity was mentioned in only 76 (out of 254) papers and, similarly, reliability was only referred to superficially in 89 papers.

Conclusion

The main goal of the study discussed in this chapter was to explore the research methodologies used in construction education, with an emphasis on mixed methods approaches. Mixed methods research is well suited to construction and civil engineering education as constructs are built using quantitative and qualitative data collection and analysis. The nature of education subjects selected for research in these fields often requires an organised and thorough investigation, which makes the mixed methods research design an effective approach. However, the general observation of methodologies currently used in the published research does not indicate a significant quota of mixed methods research designs. A collection of papers published in the *Journal of Civil Engineering* (and its previous title, *Journal of Professional Issues in Engineering Education and Practice*) from 2011 to 2021 was reviewed briefly in this paper. This journal is one of the premier references for construction and civil engineering education papers. All manuscripts published in the study range were accessed and reviewed. An initial screening was performed to limit the set of papers to just technical papers. Presumably, other types of articles did not necessarily address research projects and, therefore, were omitted from the list.

The selected papers were systematically evaluated to extract relevant patterns. Also, bibliographic information was examined from which to draw patterns. While most papers had three authors, 63% of papers did not include any undergraduate or graduate-level students. In addition, only 7% of papers were co-authored by two or more students. This indicated the relatively small size of research groups in construction management and engineering education. The type of research methodology was scrutinised in three different ways. First, a direct statement was used as a tool to categorise the papers into four groups: quantitative, qualitative, mixed methods, and others. Second, the frequency of keywords for each method was assessed. Finally, an extensive review of the methodology sections determined the type of research method used. Overall, it was determined that mixed methods approaches were used in 20% of the papers, which was an unexpectedly low number. Also, reviewing the data collection and analysis sections of all papers revealed some inaccurate or inappropriate use of statistical tools and methods. While the findings of this study indicated deficiencies and shortcomings in some research papers published, the generalisation of the results is not warranted. Despite the merits of the reviewed journal, selecting papers from other resources can alleviate this limitation. Published papers in other journals about construction and civil engineering education should be processed in the same way to attain more reliable findings. Also, increasing the duration of publication, classification based on sub-areas and locations, number of subjects used, and statistical tests can provide more insight into the research methods in construction education.

This review will help researchers in construction education to consider their research topic, context, subjects, tools, and analysis and then select a suitable research methodology. Despite the widespread use of qualitative and especially quantitative research methods in construction education research, the use of the mixed methods approach and its merits and fitness for construction education is emphasised in this chapter. The mixed methods research approach can intertwine the advantages of both qualitative and quantitative methods and provide more robust outcomes in various construction education studies.

References

Abowitz, D.A. & Toole, T.M. 2010. Mixed method research: Fundamental issues of design, validity, and reliability in construction research. *Journal of Construction Engineering and Management*, 136(1): 108–116.

Alaka, H.A., Oyedele, L.O., Owolabi, H.A., Bilal, M., Ajayi, S.O. & Akinade, O.O. 2017. Insolvency of small civil engineering firms: Critical strategic factors. *Journal of Professional Issues in Engineering Education and Practice*, 143(3): 04016026.

Bielefeldt, A.R., Lewis, J., Polmear, M., Knight, D., Canney, N. & Swan, C. 2020. Educating civil engineering students about ethics and societal impacts via co-curricular activities. *Journal of Civil Engineering Education*, 146(4): 04020007.

Bloor, M. & Wood, F. 2006. *Keywords in Qualitative Methods*. SAGE Publications Ltd., London.

Cho, C.-S., Cottrell, D.S., Mazze, C.E., Dika, S. & Woo, S. 2013. Enhancing education of construction materials course using guided inquiry modules instruction. *Journal of Professional Issues in Engineering Education and Practice*, 139(1): 27–32.

Clark, R., Spisso, A., Ketchman, K.J., Landis, A.E., Parrish, K., Mohammadiziazi, R. & Bilec, M.M. 2021. Gamifying sustainable engineering courses: Student and instructor perspectives of community, engagement, learning, and retention. *Journal of Civil Engineering Education*, 147(4): 04021009.

Creswell, J.W. 2009. *Research Design: Qualitative, Quantitative, and Mixed Methods Approaches*. 3rd edition. Sage Publications, London.

El-Diraby, T.E. & O'Connor, J.T. 2004. Lessons learned in designing research methodology in field-based construction research. *Journal of Professional Issues in Engineering Education and Practice*, 130(2): 109–114.

Flood, C. & Scott, L. 2015. *Research Methodology: A Novice Researcher's Approach*. ARCOM, Doctoral Workshop United Kingdom, Reading.

Groen, C., Simmons, D.R. & Turner, M. 2019. Developing resilience: Experiencing and managing stress in a US undergraduate construction program. *Journal of Professional Issues in Engineering Education and Practice*, 145(2): 04019002.

Holt, G.D. & Goulding, J.S. 2017. The "ological-triad": Considerations for construction management research. *Journal of Engineering, Design and Technology*, 15(03): 286–304.

Krosnick, J.A. & Presser, S. 2009. Question and questionnaire design. In: J.D. Wright & P.V. Marsden (eds). *Handbook of Survey Research*, pp. 263–313. 2nd edition. Elsevier, San Diego, CA.

Liu, J., Shahi, A., Haas, C.T., Goodrum, P. & Caldas, C.H. 2014. Validation methodologies and their impact in construction productivity research. *Journal of Construction Engineering and Management*, 140(10): 04014046.

Naaranoja, M., Kähkönen, K. & Keinänen, M. 2014. Construction projects as research objects – Different research approaches and possibilities. *Procedia - Social and Behavioral Sciences*, 119, 237–246.

Panas, A. & Pantouvakis, J.P. 2010. Evaluating research methodology in construction productivity studies. *The Built & Human Environment Review*, 3(1): 63–85.

Pandey, P. & Pandey, M.M. 2015. *Research Methodology: Tools and Techniques*. Bridge Center, Romania.

Rokooei, S., Vahedifard, F. & Belay, S. 2022. Perceptions of civil engineering and construction students toward community and infrastructure resilience. *Journal of Civil Engineering Education*, 148(1): 04021015.

Tashakkori, A. & Teddlie, C. 2010. *SAGE Handbook of Mixed Methods in Social & Behavioral Research*. SAGE Publications, Inc., London.

Zheng, L., Chen, K. & Lu, W. 2019. Bibliometric analysis of construction education research from 1982 to 2017. *Journal of Professional Issues in Engineering Education and Practice*, 145(3): 04019005.

Zou, P.X.W., Sunindijo, R.Y. & Dainty, A.R.J. 2014. A mixed methods research design for bridging the gap between research and practice in construction safety. *Safety Science*, 70: 316–326.

5 A Hybrid Project Management Model for Construction Projects

A Mixed Research Approach

Abdallah Lalmi, Gabriela Fernandes, and Souad Sassi Boudemagh

Summary

In this chapter, a hybrid project management (PM) model for the construction industry, which combines traditional PM with agile and lean approaches, is presented. The purpose of the model is to promote flexibility and stimulate interaction and collaboration between stakeholders, to increase project success. A sequential explanatory mixed methods research design was used for the study. First, a questionnaire survey was conducted to discover how frequently traditional, agile, and lean management practices are employed. Second, based on the literature review and the questionnaire results, we formulated an initial, proposed hybrid model. Third, a qualitative explanatory case study, using a virtual focus group, was carried out to improve and validate the model. The Hybrid Management Model was conceptualised as a basis on which a construction organisation can adapt better to its context. The main contribution is a demonstration of the use of the explanatory sequential mixed methods research design in the development of a hybrid PM model for construction projects, which relies on the interaction of practices from the three different management approaches. The findings of the study were based on a single case study of a large-scale construction project of an Algerian tram system. Future studies might involve multiple case studies for cross-checking the conclusions.

Introduction

Researchers have recognised the adoption of PM as a critical factor in the success of construction projects (Demir *et al.*, 2013). However, the true value of PM is a function of how it is implemented and how well it fits into the organisational context (Cooke-Davies *et al.*, 2009). Choosing the right PM methodology is key to project success (Fernandes & O'sullivan, 2023; Fernandes *et al.*, 2018; Lalmi *et al.*, 2021). Hybrid PM can increase the chances of project success in the construction sector (Lalmi *et al.*, 2022). It emerges from the desire to realise the benefits of Agile methodologies while retaining some of the structure of so-called "Waterfall" methodologies (Špundak, 2014). Schmitz *et al.* (2018) explained that to adapt specific methodologies to a project, Hybrid PM is used to borrow, mix, and combine processes from both Agile and plan-based methods.

DOI: 10.1201/9781003204046-5

However, in this research, the notion of Hybrid concerns the combination of three approaches: traditional, agile, and lean. The aim of Traditional PM practices is to maintain the value of a predictable project, while Agile management practices allow for a higher degree of interaction with the client and flexibility in response to change. The reduction of non-value-added activities and waste in projects is advocated in Lean PM. The use of Traditional, Agile, and Lean practices is expected to lead to better customer satisfaction, eliminate waste, and ensure a high probability of the success of a construction project. PM practices are the mechanisms by which PM processes are delivered and supported and which, when managed effectively, can lead to project success (Barbosa *et al.*, 2021).

Construction projects are often large and complex; and are usually characterised by several uncertainties that are associated with them, hence difficult to manage. Construction projects have the following attributes: they are high-tech, capital intensive, and involve engineering projects of such a significant scale that companies must work collaboratively across functions to complete (Sacks *et al.*, 2015).

As a result of the difficulties and challenges posed by Traditional PM methods, especially the management of uncertainty and change, a tendency towards project failure has been observed in the construction industry (Mahamid *et al.*, 2012). Agile PM is a PM approach that responds well to change (Denning, 2013). Indeed high effectiveness in applying Agile PM has been demonstrated in PM in general (Davidson & Klemme, 2016). Hard and soft approaches should not be opposites but, instead, a mixture of the two could create a PM style that fits any specific project better (Gustavsson & Hallin, 2014).

To benefit from the best practices of each approach, a robust new PM model that allows for increased complexity and stakeholder involvement is needed to deliver a project on time, within budget, and without significant scope creep. The construction industry should adapt to more agile techniques to ensure long-term success (Al Behairi, 2016). For a project to succeed, managers must ensure that their choices and actions will have a positive effect on the outcome of the project and, consequently, on the profit of the company. In deciding how to proceed with a project, managers must be sure that the chosen model will optimise the chances of project success, thus contributing to the achievement of the organisation's strategic objectives. Thus, the main question of this research emerged: How can Traditional, Lean, and Agile approaches be integrated to improve the likelihood of success in construction projects?

To address this research question, the aim of the researchers was to develop a new Hybrid PM model for the construction industry, which includes various practices from the three approaches: Traditional, Agile, and Lean, by implementing a mixed method research design. Combining both qualitative and quantitative methods made it possible for the data to be triangulated, but also allowed for the complementarity of the different aspects studied (Saunders *et al.*, 2019), especially the relationships between the different quantitative variables used in the questionnaire, to develop a Hybrid PM model geared specifically to construction projects.

By employing a sequential, explanatory, mixed methods research design, knowledge was developed during this research in the area of PM by adopting a practical process view for developing a holistic and structured PM model for construction projects. The study commenced with the development of a questionnaire survey, then a Hybrid PM model was built based on the findings of a literature review and the results of a questionnaire survey. The model was improved based on input from key actors involved in a major project for an Algerian tram system and was validated by engaging with a focus group consisting of key experts from different organisations involved in a major construction project.

The Hybrid model involved applying not only the commonly-used Traditional practices but also Agile and Lean practices to produce results faster, more cheaply, and more effectively. This was the first model in the construction industry in which the practices of three different management approaches were combined, bringing important practical contributions to the field, particularly for project managers and other key management stakeholders involved in construction projects.

Hybrid PM Approach in Construction Projects

Gemino *et al.* (2021) argued that the combination of Traditional and Agile approaches leads to project success. Findings by various researchers suggest that Hybrid PM is becoming increasingly necessary (Gemino *et al.*, 2021; Fernandes & O'Sullivan, 2023). Al Behairi (2016) noted that the boundaries between Traditional and Agile projects have become blurred and recognised that a Hybrid model is not a new practice, but merely the re-application of existing practices in new ways. Al Behairi (2016) provided a sound foundation for why a Hybrid between Traditional and Agile PM techniques is needed for the construction industry.

The introduction of a new Hybrid approach and its subsequent acceptance by a project team is necessary before beginning PM activities. Converting or introducing a Hybrid approach in the middle of the life cycle of a project is unwise and increases the probability of project failure. Furthermore, it would be necessary to agree upon any Hybrid involving Agile PM with the client beforehand, as this practice relies heavily on client involvement (Nikravan & Melanson, 2008). Risk assessment, stakeholder management, and other methods prescribed by Archer and Kaufman (2013) remain valid activities but might be redundant or unnecessary in a truly Agile approach, which ensures customer engagement, and a Lean approach, which reduces waste and eliminates non-value-added activities (Demir *et al.*, 2013). Tarne (2015) prescribed that, regardless of the PM model chosen, a clear understanding of the project vision is needed from the start, and that an unambiguous communication plan should be available consistently across all project methodologies.

It has been shown that the appropriate choice of PM methodology or model, with the relevant set of PM practices, is positively related to project success (Badewi, 2016; Joslin & Müller, 2016). The goal of the Hybrid approach is to retain the predictability of a Traditional approach while allowing for greater agility. All the functions of a Hybrid approach should work together to become more Agile (Cobb, 2012).

In many of the extant comparisons between Agile and Traditional projects, Agile ideals are compared with a poorly managed Traditional project, but the three approaches, Traditional, Agile, and Lean, specifically proposed in this study are addressed in only a limited amount of research. Even though Traditional methods are shown in a negative light in some comparisons, many of these practices are still valid in a Hybrid model (Cobb, 2012), especially in the context of a construction project because its planning is still based on a Traditional approach. The most important concept for any Hybrid approach is the understanding that flexibility is the key to a good model.

Gemino *et al.* (2021) found that the effectiveness of Hybrid approaches has been shown to be similar to wholly Agile approaches. Practitioners suggest a combination of Agile and Traditional practices and believe that a Hybrid approach is a leading PM approach. Although Hybrid models are discussed in the literature in which combinations of Traditional and Agile approaches are proposed, there are no studies in which the impact of adopting this Hybrid approach in organisations in practice is addressed (Azenha *et al.*, 2021). For construction projects, it is necessary for any Hybrid approach to be well conceived, using the best practices of each approach, while discarding techniques and tools that have proven to be ineffective or outdated based on the literature. By using the strengths of the three approaches – Traditional, Agile, and Lean – the aim of a Hybrid model is to mitigate the effects of the causes of failure, as observed in construction projects, and strive to achieve cost-effective projects in the construction industry by means of effective PM practices and dynamic, collaborative environments.

Research Methodology

The choice of research design is crucial to the success and effectiveness of any research study. Commonly, in the social sciences, research methodology is clustered around (1) deductive approaches using quantitative methods (quantitative research), or (2) inductive approaches using qualitative methods (qualitative research). However, contributions can also be made when (3) quantitative methods are used to lay the ground for inductive theory building or (4) qualitative methods are used to test deductive theory (Saunders *et al.*, 2019). Another alternative is the mixing of methods, which can be controversial (Saunders *et al.*, 2019). Some researchers believe that the assumptions underlying qualitative and quantitative approaches are inherently opposed and, therefore, that the two cannot be combined meaningfully. Others believe that the two approaches are best used in combination only by alternating between methods (Saunders *et al.*, 2019).

In this study, there were two main reasons for using a sequential, explanatory, mixed research design: data triangulation and complementarity (Saunders *et al.*, 2019). Triangulation was done by using independent sources of data to corroborate findings within the study. Complementarity was achieved by studying a variety of different aspects, especially, the relationships between the quantitative variables to develop a Hybrid PM model.

Quantitative Approach

The online questionnaire survey was conducted to allow for an easy comparison of responses across different locations or periods (Saunders *et al.*, 2019).

The questionnaire was divided into three parts. The first part included general information about the respondents, while the second and third parts addressed the most frequently used and beneficial practices in the Traditional, Agile, and Lean approaches. These practices are related to the different phases of the project life-cycle: initiation, planning, execution, monitoring and control, and closure. The choice of techniques and tools based on cross-referencing the published literature and studies by different authors applied in the construction industry (Besner & Hobbs, 2006; Papke-Shields *et al.*, 2010; Fernandes *et al.*, 2013; Tereso *et al.*, 2019). The survey addressed 34 Traditional practices, 9 Agile practices, and 13 Lean construction practices. The respondents were asked to rank the degree of use of each technique and tool, using a scale of 1–5 where: 1 = "less-used", 2 = "rarely used", 3 = "occasionally used", 4 = "often used", and 5 = "always used".

The questionnaire was distributed among the researchers' professional contacts (project managers, programme managers, and project management officers (PMOs). The lack of accurate information and accessibility of information were the reasons for choosing a non-probability sampling method (Saunders *et al.*, 2019). Data collection was conducted from March 2020 until September 2020. Data collected was analysed using the SPSS software package, and descriptive/inferential statistics were applied for data analysis.

Qualitative Approach

To improve and validate the initial, proposed Hybrid PM model, a virtual focus group discussion was held using the Zoom platform.

Six key experts were selected based on their role and experience in managing construction projects and included project managers, PMOs and project team members. All participants had extensive experience (more than 15 years) in managing construction projects. A virtual focus group was used instead of an in-person group because of the geographical separation of the experts (Bloor, 2001). The focus group meeting was conducted without a rigid structure, so the discussion was free-flowing to enable comprehensive discussion of the main proposed project phases, the different Traditional, Agile and Lean practices, and the relationships between these practices based on the results of the questionnaire survey and literature review, as depicted in the proposed Hybrid model. The discussion was guided by the questions: "What are the Traditional/Agile/Lean practices often applied across your delivery process?" and "What are the challenges commonly faced in adopting these practices?", and content analysis was applied.

The validity and reliability of the data used in the Hybrid model were tested using the "reader-author" cycle, that is the researchers created various drafts as "authors" of the model and then passed these on to multiple participants. The latter, as "readers" then validated the accuracy of the model. Therefore, the Hybrid model was developed collaboratively between the researchers and the six experts involved in the focus group.

Results and Discussion

Most Used Traditional, Lean, and Agile Practices

Table 5.1 shows the reliability analysis of the questionnaire data. Cronbach's alpha varied between 0.830 and 0.938, all above the 0.5 minimum threshold, and the 0.7

desired threshold, which meant that the results are reliable (Saunders *et al.*, 2019). Table 5.2 shows the descriptive statistics and ranking of the most frequently used Traditional, Agile, and Lean practices.

The relative importance index (RII) and the mean items score (MIS) were calculated using the mean of responses:

$$RII = \frac{(\Sigma an)}{N} \tag{5.1}$$

and

$$MIS = \frac{(\Sigma an)}{N} \tag{5.2}$$

where

"a" is a constant that expresses the weight given to each practice, which varied from 1 to 5;
"n" is the response frequency;
"N" is the total number of respondents.

The standard deviation is the square root of the variance, and it measures the dispersion of the dataset relative to its mean. If two practices had the same mean, the practice with the lowest standard deviation was classified first. Therefore, using the RII and the standard deviation, the practices were classified in descending order.

Results from the questionnaire suggested that the ten most commonly used practices were: Kick-off meeting, ranked first as the most widely used practice in terms of frequency (RII (%) = 76.4), which was the same as the result obtained in previous research (Perrotta *et al.*, 2017); Teamwork (RII (%) = 74.8); Progress meeting (RII (%) = 74.2); Progress reports (RII (%) = 72.6); Contract closure (RII (%) = 73.6); Project close-out report (RII (%) = 71.6); Project close-out meeting (RII (%) = 69.8); Lessons learned (RII (%) = 69); WBS (RII (%) = 68.4); and Project charter (RII (%) = 68.4).

The integration of Agile practices into the construction industry was still a novelty. Most Agile practices were used in the context of Information Technology. The ranking of the practices was as follows: Daily meeting (RII (%) = 56.4); Kanban board (RII (%) = 53.6); Incremental planning (RII (%) = 52.2); Sprint

Table 5.1 Reliability analysis

Practices	*No of items*	*Cronbach's alpha*
Traditional practices	34	0.938
Agile practices	9	0.900
Lean practices	13	0.830

Source: Original.

Table 5.2 Descriptive statistics and ranking of the most frequently used traditional, agile, and lean practices

Approach	*Practices*	*Ranking*	*Mean*	*Std deviation*
Traditional practices	Kick-off meeting	1	3.82	1.136
	Teamwork	2	3.74	1.178
	Progress meeting	3	3.71	1.313
	Closing of contracts	4	3.68	1.317
	Progress report	5	3.63	1.364
	Project closure report	6	3.58	1.407
	Project closing meeting	7	3.49	1.465
	Lessons learned	8	3.45	1.519
	WBS	9	3.42	1.348
	The project charter	10	3.42	1.518
	Project management plan	11	3.34	1.361
	Issue log	12	3.21	1.527
	Change log	13	3.18	1.574
	Risk register	14	3.11	1.467
	Periodic audit control	15	3.08	1.383
	Business case	16	3.08	1.496
	Stakeholders register	17	3.05	1.355
	The critical path	18	2.95	1.627
	Earned schedule management	19	2.92	1.282
	Earned duration management	20	2.87	1.212
	Statistical control charts	21	2.79	1.417
	Earned value management	22	2.79	1.492
	Earned value analysis	23	2.76	1.478
	Resource buffering	24	2.71	1.469
	Project management information system	25	2.71	1.541
	PERT method	26	2.58	1.287
	Building information modelling (BIM)	27	2.58	1.368
	Allocation approaches under uncertainty	28	2.53	1.370
	Matrix-based method	29	2.42	1.348
	Temporary structure-planning generator	30	2.37	1.125
	Monte Carlo simulation	31	2.32	1.435
	Facility location model	32	2.16	1.151
	Fuzzy decision node	33	2.16	1.151
	Rough-cut capacity planning	34	2.08	1.148
Agile practices	Daily meeting	1	2.82	1.373
	Kanban board	2	2.68	1.416
	Incremental planning	3	2.61	1.462
	Sprint retrospective	4	2.55	1.329
	Cycle planning	5	2.50	1.502
	Sprint planning	6	2.47	1.310
	Sprint review	7	2.42	1.328
	User story	8	2.37	1.303
	Moscow technique	9	2.21	1.379

(*Continued*)

Table 5.2 (Continued)

Approach	*Practices*	*Ranking*	*Mean*	*Std deviation*
Lean practices	Teamwork and partnership	1	2.63	1.460
	Kanban	2	2.61	1.480
	Just in time	3	2.58	1.407
	Virtual design construction	4	2.58	1.464
	Total quality management	5	2.58	1.426
	Visualisation tools	6	2.47	1.330
	Target value design	7	2.42	1.388
	Last planner system	8	2.39	1.462
	Integrated project delivery	9	2.39	1.405
	Fail-safe for quality and safety	10	2.34	1.279
	Total productive/Preventive maintenance	11	2.34	1.419
	BIM	12	2.32	1.276
	The plan for working conditions and environment in the construction industry	13	2.32	1.358

Source: Original.

retrospective (RII (%) = 51); Cycle planning (R (%) = 50); Sprint planning (RII (%) = 49.4); Sprint review (RII (%) = 48.4); User story (RII (%) = 47.4); and Moscow technique (RII (%) = 44.2).

Lean construction practices were only recently being introduced to construction projects and their use remained limited. Many professionals in the construction sector were simply not aware of the added value of Lean. A set of techniques and tools was ranked, based on the respondents' answers: Teamwork and partnership (RII (%) = 52.6); Kanban (RII (%) = 52.2); Just-in-time and Virtual design construction (RII (%) = 51.6); Total quality management (RII (%) = 52.2); Visualisation tools (RII (%) = 49.4); Target value design (RII (%) = 48.4); and Last Planner System (RII (%) = 47.8).

Initial Proposed Hybrid Model

The initial, proposed Hybrid model, presented in Figure 5.1, illustrates the suggested Hybrid PM approach for construction projects, based on the results of the questionnaire survey and literature review. The main objective of this Hybrid model is to reduce the overall cost and time-frames when compared with the more commonly-used Traditional practices in construction projects. Traditional practices are used in an attempt to control the scope by having an appropriate level of detailed planning (PMI, 2021). The aim of Agile practices is to involve the customer and ensure the latter's engagement throughout the sprints in the design phase, thus guaranteeing collaboration, especially during prioritisation of customer requirements (Archer & Kaufman, 2013). The aim of Lean practices is to reduce waste and eliminate waiting time in the design phase and re-planning, by using the Last Planner System, for example (Demir *et al.*, 2013). When this Hybrid model is applied, closer collaboration with the customer is required.

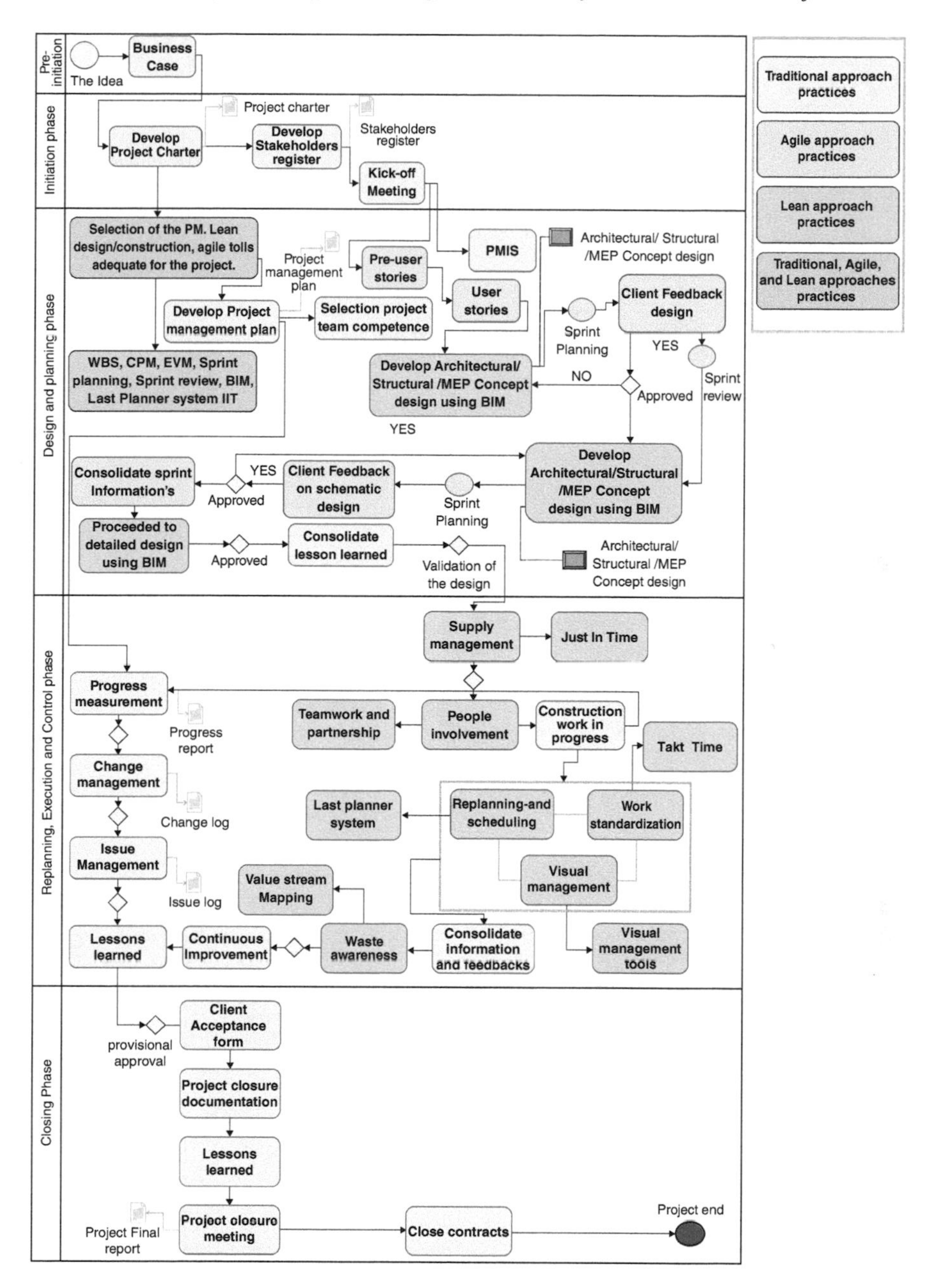

Figure 5.1 Initial proposed hybrid model for construction projects

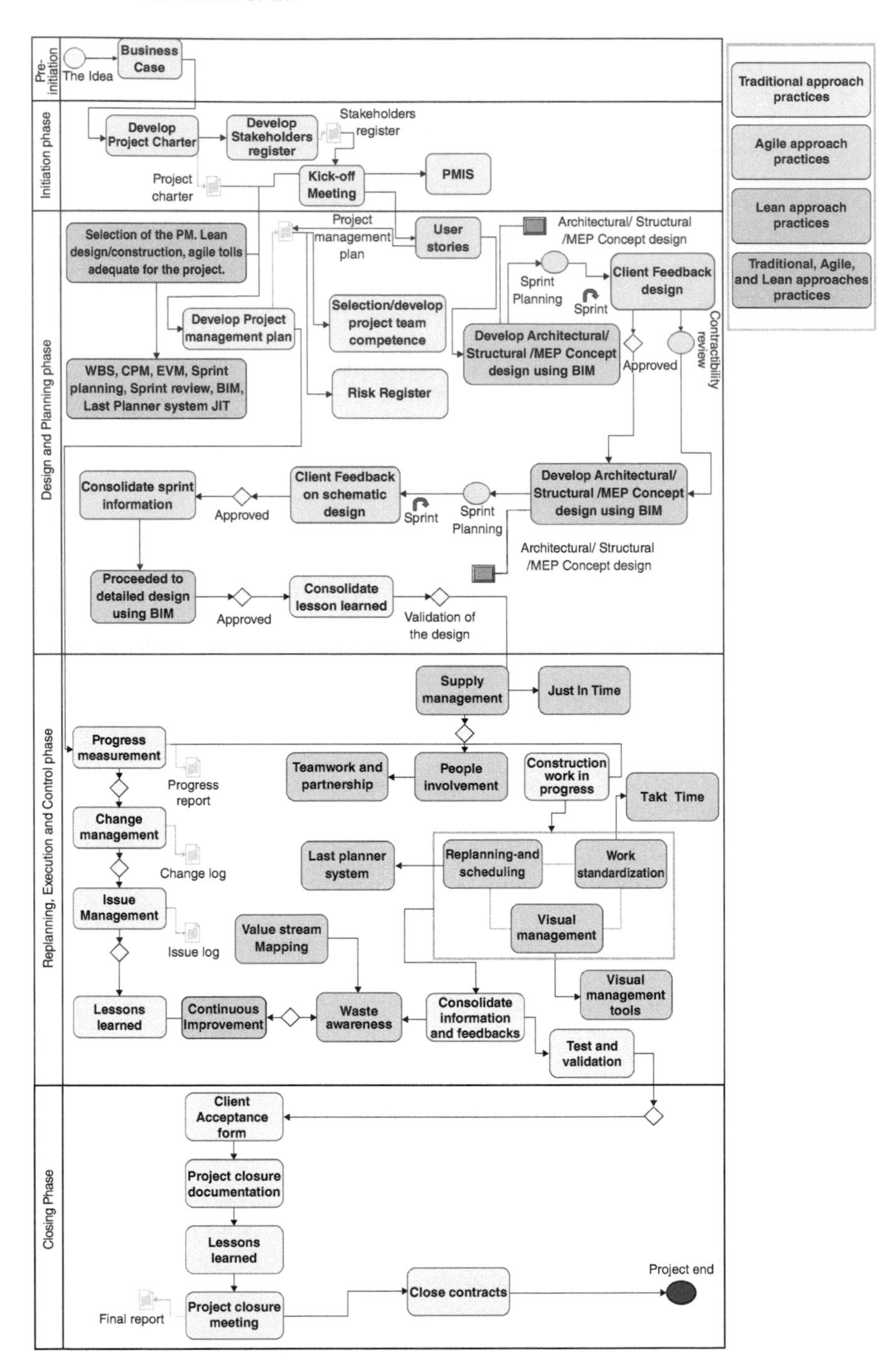

Figure 5.2 Hybrid PM model for construction projects

Improved Hybrid Model

Improvement and validation of the model was the last step of this research. Although a focus group is not considered to be the best research method for validating results, in practice it was found to be the most appropriate research method bearing in mind the complexity of the artefact (Figure 5.1) that was being validated. A focus group of six experienced experts in construction PM was conducted by the researchers to improve and validate the Hybrid model. The participants agreed on the proposed project phases, although several remarks were noted.

Concerning the "Initiation phase", some changes to links were considered to be necessary: the link runs from the "Stakeholder register" to the "Kick-off meeting". Most of the experts judged that the "Project management information system" should be implemented at the beginning of the project, that is before the launch of the calls for tender. It was also better if it is linked immediately after the "Project charter".

In the "Planning and design phase", the experts agreed on the lack of a "Risk register", which is highlighted in the literature (Tereso *et al.*, 2019) and, therefore, was added. The "User story" is formally expressed as an Agile practice, at the level of the design phase, and the expert participants noted the lack of a sprint operation in the design phase, with some of them proposing to substitute "sprint" for "release" and "sprint review" for "contractibility review". Some of them also noted that there would be no use for "no and yes" options since this study was concerned with Lean and, given that, among the purposes of Lean, is the elimination of waste and non-value-added elements, the use of BIM as a tool for Lean design reduces the waiting time between different stakeholders. Changes were required to the linkages between practices to ensure the flexibility of the proposed model. The experts recommended adding the "Kick-off meeting" at the end of the planning phase, as suggested in PMBoK (PMI, 2021).

In the "Execution and control phase", the major changes made were a change of the link between "Lessons learned" and "Provisional acceptance" by adding another new practice, "Test and validation" before moving on to provisional acceptance; the change of location between "Value stream mapping" and "Waste awareness" – it was necessary first to map this awareness of waste to occur, which then leads to "Continuous improvement". Actually, it was already a common practice because it is integrated into all three approaches. Regarding "Project closure", all the focus group participants agreed on the selected practices, except for the suggestion that "Lessons learned" be replaced by "Consolidate lessons learned" because, in the final phase, all the lessons previously collected are gathered and stored.

The aim of the final, proposed Hybrid PM model, presented in Figure 5.2, is to manage a project better by reducing the conditions for failure within the framework of a well-defined yet flexible model, thereby ensuring that the results of this research can be exploited effectively and effected swiftly at the organisational level.

Implications for Practice and Research

This research makes theoretical, practical, and methodological contributions. First, it extends existing knowledge on PM by adopting a practical process view for developing a holistic and structured PM model for construction projects, bringing a micro-level perspective to PM. Second, the research improves the theoretical understanding of how to combine three different approaches in the management of projects: Traditional, Agile and Lean, with a view to assuring overall project success. In the chapter, a Hybrid PM model proposal to manage construction projects is presented. The current knowledge on key Traditional, Agile and Lean practices in the particular context of construction projects is articulated in the model, which can also guide projects in other contexts in their adoption of Hybrid PM approaches.

In this chapter, an important contribution is made to practice by emphasising the need to combine different PM approaches to deal with the increasing complexity of construction projects (Al Behairi, 2016). This implies the development of a formal and unique PM model for the context of each particular construction project, for which the developed Hybrid PM model might be used as a starting point.

Finally, in this chapter, empirical evidence is also brought to light on the importance of using sequential, explanatory, mixed methods research designs in PM studies, especially for data triangulation and complementarity, and other reasons, for example facilitation, generality, and to aid interpretation or solving a problem (Saunders *et al.*, 2019). Additionally, the importance of engaged and collaborative research is reinforced, that is between academics and practitioners, as both parties play an active role in PM studies, to be able to propose more robust theories that lead to better project performance in practice (Brunet, 2022).

Conclusions

The complexity of construction projects has increased, leading to rising costs, missed deadlines, unverifiable changes, and site waste. However, the use of the proposed Hybrid model would enable a project team to reduce the duration and cost of a project, with an emphasis on greater client interaction and collaboration and the elimination of non-value-added elements.

The implementation of the Hybrid management approach in construction projects was reviewed in the chapter by discussing the strengths and weaknesses of Traditional, Agile, and Lean PM approaches and the best practices applicable to construction projects, to propose a PM model which can overcome the common failure modes observed in construction projects. The development of the proposed model involved an analysis of extant synergies between the practices associated with the three management approaches to establish the complementarity between them and to see the possibility and practicality of introducing this new model, which is still considered to be a novelty within the construction industry context. In the Hybrid model, some critical PM practices specific to the construction projects are identified, such as "Building Information Modelling". However, other essential PM practices in the Hybrid model are common or generic to any project, including the "project charter", "kick-off meeting" or "progress reports". In this research, the use of sequential, explanatory, mixed methods research design enabled the researchers to gain in-depth knowledge of the subject under study.

Certain drawbacks of the research study, resulting mainly from the decisions concerning the methodological approach, are acknowledged. First, the period for data collection from questionnaire survey was limited to three months. Second, the emergent Hybrid model was based on literature review, a questionnaire survey, and a case study using a focus group. Therefore, this Hybrid model cannot be generalised for all construction project contexts, nor can it be used for other industry contexts. In this regard, future studies may include multiple case studies to validate the conclusions among them, thereby increasing the generalisability of the results.

References

Al Behairi, T.A. 2016. *AGISTRUCT: Improved Model for Agile Construction Project Management. Paper presented at PMI® Global Congress 2016—EMEA, Barcelona, Spain.* Project Management Institute, Newtown Square, PA.

Archer, S. & Kaufman, C. 2013. Accelerating outcomes with a hybrid approach within a waterfall environment. Available at: http://www.pmi.org.

Azenha, F.C., Reis, D.A. & Fleury, A.L. 2021. The role and characteristics of hybrid approaches to project management in the development of technology-based products and services. *Project Management Journal*, 52(1): 90–110.

Badewi, A. 2016. The impact of project management (PM) and benefits management (BM) practices on project success: Towards developing a project benefits governance framework. *International Journal of Project Management*, 34(4): 761–778.

Barbosa, A.P., Salerno, M.S., de Souza Nascimento, P.T., Albala, A., Maranzato, F.P. & Tamoschus, D. 2021. Configurations of project management practices to enhance the performance of open innovation R&D projects. *International Journal of Project Management*, 39(2): 128–138.

Besner, C. & Hobbs, B. 2006. The perceived value and potential contribution of project management practices to project success. *Project Management Journal*, 37(3): 37–48.

Bloor, M. 2001. F*ocus Groups in Social Research*. Sage, London.

Brunet, M. 2022. On the relevance of theory and practice in project studies. *International Journal of Project Management*, 40(1): 22–24.

Cobb, C.G. 2012. En désaccord ? Réseau des PM. Available at: www.pmi.org

Cooke-Davies, T.J., Crawford, L.H. & Lechler, T.G. 2009. Project management systems: Moving project management from an operational to a strategic discipline. *Project Management Journal*, 40(1): 110–123.

Davidson, A. & Klemme, L. 2016. Why a CEO should think like a scrum master. *Strategy & Leadership*, 44(1): 36–40. https://doi.org/10.1108/SL-11-2015-0086

Demir, S.D., Bryde, D.J. & Sertyesilisik, B. 2013. Introducing AgiLean to construction project management. *Journal of Modern Project Management*, 1(3): 1–18.

Denning, S. 2013. Why Agile can be a game changer for managing continuous innovation in many industries. *Strategy & Leadership,* 41(2): 5–11.

Fernandes, G., Moreira, S., Araújo, M., Pinto, E.B. & Machado, R.J. 2018. Project management practices for collaborative university-industry R&D: A hybrid approach. *Procedia Computer Science*, 138: 805–814.

Fernandes, G. & O'Sullivan, D. 2023 Project management practices in major university-industry R&D collaboration programs–a case study. *The Journal of Technology Transfer*, 48(1): 361–391.

Fernandes, G., Ward, S. & Araújo, M. 2013. Identifying useful project management practices: A mixed methodology approach. *International Journal of Information Systems and Project Management*, 1(4): 5–21.

Gemino, A., Horner Reich, B. & Serrador, P.M. 2021. Agile, traditional, and hybrid approaches to project success: Is hybrid a poor second choice? *Project Management Journal*, 52(2): 161–175.

Gustavsson, T.K. & Hallin, A. 2014. Rethinking dichotomization: A critical perspective on the use of "hard" and "soft" in project management research. *International Journal of Project Management*, 32(4): 568–577.

Joslin, R. & Müller, R. 2016. Identifying interesting project phenomena using philosophical and methodological triangulation. *International Journal of Project Management*, 34(6): 1043–1056.

Lalmi, A., Fernandes, G. & Boudemagh, S. 2022. Synergy between traditional, agile and lean management approaches in construction projects: Bibliometric analysis. *Procedia Computer Science*, 196: 732–739.

Lalmi, A., Fernandes, G. & Souad, S.B. 2021. A conceptual hybrid project management model for construction projects. *Procedia Computer Science*, 181: 921–930.

Mahamid, I., Bruland, A. & Dmaidi, N. 2012. Causes of delay in road construction projects. *Journal of Management in Engineering*, 28(3): 300–310.

Nikravan, B. & Melanson, D. 2008. Application of hybrid agile project management methods to a mission-critical law enforcement agency program. Available at: http://www.pmi.org

Papke-Shields, K.E., Beise, C. & Quan, J. 2010. Do project managers practice what they preach, and does it matter to project success? *International Journal of Project Management*, 28(7): 650–662.

Perrotta, D., Fernandes, G., Araújo, M., Tereso, A. & Faria, J. 2017. *Usefullness of Project Management Practices in Industrialization Projects – A Case Study. Proceedings of 2017 International Conference on Engineering, Technology and Innovation (ICE/ITMC)*, Madeira Portugal, 1104–1112.

PMI, 2021. *PMBOK-Guide to the Project Management Body of Knowledge*. 6th edition. Pennsylvania, USA.

Sacks, R., Whyte, J., Swissa, D., Raviv, G., Zhou, W. & Shapira, A. 2015. Safety by design: Dialogues between designers and builders using virtual reality. *Construction Management and Economics*, 33(1): 55–72.

Saunders, M.N.K., Lewis, P. & Thornhill, A. 2019. *Research Methods for Business Students*. 8th edition. Pearson Professional Limited. Harlow.

Schmitz, K., Mahapatra, R. & Nerur, S. 2018. User engagement in the era of hybrid agile methodology. *IEEE Software*, 36(4): 32–40.

Špundak, M. 2014. Mixed agile/traditional project management methodology – Reality or illusion? *Procedia-Social and Behavioral Sciences*, 119: 939–948.

Tarne, R. 2015. *Why agile may not be the silver bullet you're looking for*. PMI Global Congress, Orlando, FL.

Tereso, A., Ribeiro, P., Fernandes, G., Loureiro, I. & Ferreira, M. 2019. Project management practices in private organizations. *Project Management Journal*, 50(1): 6–22.

6 Enhancing the Employability of Quantity Surveying Graduates

A Mixed Methods Approach

Samuel Adekunle, Iniobong Beauty John, Obuks Ejohwomu, Clinton Aigbavboa, Andrew Ebekozien, and Douglas Aghimien

Summary

Employability has been established to have far-reaching implications for the success of individual careers and firms. However, there is still a dearth of employability studies in quantity surveying (QS) in which a mixed methods research design is used. The aim of this study was to bridge this gap. Quantitative data were sourced using pre-formatted questionnaires which were administered to industry professionals. Also, semi-structured interviews were conducted with principal partners, associate partners, and staff members in charge of recruitment in QS firms situated in Lagos, Nigeria. The adoption of a sequential explanatory mixed methods approach provided a more robust and deep insight into the employability of QS graduates. Seventeen competencies were highlighted in the study. Based on the study, it was concluded that a wide gap exists between the observed competencies (for instance, trustworthiness and positive attitude) and expected competencies (for instance trustworthiness and effective communication). In addition, employers prefer specific behavioural skills in graduates as most believe that professional competencies can be developed on the job. It was proposed that the existing curriculum be re-working, among other solutions towards bridging the gap between industry needs and the skills observed among QS graduates. The purpose of this study was to contribute to the existing literature about the employability of graduates and to outline the utility of adopting a mixed methods research design.

Introduction

The construction industry is undergoing a rapid transformation because of technological advances. Technological innovation gave rise to disruptions throughout the industry – people, processes, projects, and organisations. The observed disruption is not stakeholder- or project-specific, and it introduced new expectations and deliverables.

DOI: 10.1201/9781003204046-6

The dynamic nature of the workplace has influenced the skill set required of graduates from tertiary education institutions. For instance, Adekunle *et al.* (2019) noted that a gap exists between the competencies expected by employers and the competencies acquired by graduates from tertiary institutions. Evidence suggests that employers in the construction industry are dissatisfied with the technical skills acquired by graduates (Aliu & Aigbavboa, 2020). Also, employers complain of graduates not having competencies that meet their expectations (Davies *et al.*, 1999). The employability of graduates has been considered to be inadequate by employers (Anho, 2011). Consequently, the employability of graduates must be enhanced to meet the requirements of the dynamic nature of the industry. By extension, the following must also be prioritised to achieve alignment with the industry requirements: the learning approaches used in the education system (Ornellas *et al.*, 2018), and re-working the curriculum (Su & Zhang, 2015; Adekunle *et al.*, 2019), among others.

On the other hand, evidence suggests that the programmes of some institutions of learning meet industry expectations (Mansour & Dean, 2016). In the final analysis, it is obvious that the outputs of tertiary institutions must be tailored to meet industry expectations. This is because employability competencies act as a stimulant for career success (Van Der Heijde & Van Der Heijden, 2006). Competencies refer to a set of skills and characteristics that enhances job performance. Therefore, competencies remain a determinant of employability; they are acquired and determine job fit. According to Srinath and John (2013), the Royal Institute of Chartered Surveyors (RICS) divides the competencies of a quantity surveyor (QS) into the following three types:

- **Mandatory competencies**: are generally required by the surveying professions. These are the basic skill sets necessary for working as a professional in the construction industry.
- **Core competencies:** represent the core skills base of QSs and are essential for practising as a QS.
- **Optional competencies**: are competencies desired in a QS.

The three categories of competencies at the graduate level are critical for providing a foundation level of knowledge. Similarly, the Australian Institute of Quantity Surveyors also has an outline of essential abilities that define a competent quantity surveyor, as presented in Table 6.1.

Table 6.1 Fundamental competencies required of a quantity surveyor

Competencies	*Explanation*
Quantification/ measurement.	The ability to: i understand and apply the standard method of measurement relevant to the area of practice ii understand and use standard phraseology of building trades and elements iii quantify, enumerate, and measure.
Communication skills	The ability to: i communicate effectively, orally, in writing, and with visual aids ii combine facts or ideas into a complex whole iii prepare written information in a formal way that conveys meaning.
Personal and interpersonal skills	The ability to: i demonstrate self-confidence, time management, self-motivation, and enthusiasm ii understand the role and motivation of others and participate in professional and inter-professional teamwork iii identify and assess problems and find innovative solutions iv set and achieve personal objectives and targets v understand and, where appropriate, apply marketing and negotiating skills.
Business and management skills	The ability to: i recognise the need for cost-effective use of appropriate resources ii understand the process of quality control and assurance, and understand the suitable certification iii identify consumer and client needs and the process for their satisfaction iv understand accounting principles, including budgets and cash flows v understand the scale of fees and charges for professional services vi be familiar with general economic principles.
Professional practice	The ability to: i recognise the nature and significance of property development in all its forms ii understand the role responsibilities and legal liabilities of quantity surveyors in matters of practice iii understand and apply the ethics of professional practice iv understand and apply legislation relevant to providing a professional service, including registration of quantity surveyors and quantity surveying practices v understand the role of quantity surveyor in a multi-disciplinary project team vi understand the structure of the National Institute of Quantity Surveyors, its by-laws and rules of conduct.
Computer and information technology	The ability to: i understand and apply basic computer skills relevant to the area of practice ii understand the use and relevance of information technology

Table 6.1 (Continued)

Competencies	*Explanation*
Construction technology	Acquire knowledge of: i Construction processes and technologies ii Construction activities and sequencing of activities iii Source and use of building materials, including testing and assessing techniques iv Design and installation of building services v Principles of building science in respect of heat, light, and sound vi Principles of building science in relation to structures including analysis, design, and stability vii Principles of construction including demolition methods, formwork design, erection techniques, plant and equipment viii Principles of site surveys ix Interpretation of building plans, construction codes, and regulations x Specification writing
Construction law and regulation	Acquire knowledge of: i laws and regulations relevant to the construction industry ii various forms of building and construction contracts.

Source: AIQS (2012).

Research Methodology

Mixed methods research methodology is observed to be supported by a pragmatic philosophical paradigm. An in-depth literature review was undertaken to analyse pragmatism as the research philosophy backing mixed methods research, as Maarouf (2019) suggested that it provides a framework within which to explore all possible options available to the researcher. Pragmatism thus offers a robust perspective for achieving a research study. The rest of this chapter contains details about the adoption of a mixed methods approach towards understanding the employability of QS graduates.

Research Approach

The aim of every research effort is to achieve an objective and contribute to an existing body of knowledge. Other reasons for conducting research include solving a problem and propounding a theory. However, to accomplish this, an enquiry approach must be adopted in research endeavours to unravel the research problem. This can either be a singular or multiple enquiry method(s). The adoption of a mixed methods research approach entails using more than one enquiry approach to achieve research objectives. The advantage of this approach is that the methods adopted complement each other (in terms of weakness and strength). The approach also improves the validation and robustness of the result. The mixed methods research approach enables researchers to incorporate multiple research approaches in a study, thus presenting a rigorous and thorough enquiry method. It can also provide a basis for comparative analysis and understanding of a concept.

The mixed methods approach involves the combination of different research approaches to achieve the aim of a study. The mixture can be in terms of theories, data, investigators, and methodologies (Cameron & Miller, 2007). A combination of quantitative and qualitative data collection methods (a questionnaire survey and interviews) was adopted in this study. This methodological choice was considered to be appropriate to provide a robust understanding of the problem being studied. Specifically, a sequential explanatory mixed methods variant was adopted. This was to increase the validity and robustness of the findings of the study. This was achieved based on the cross-validation of datasets that emerged from the adoption of different data collection methods. Figure 6.1 shows the sequential explanatory mixed methods research design adopted for the study.

Quantitative and qualitative research methods were adopted in this study to answer the research questions objectively. The mixed methods approach has been adopted in existing studies to understand better the phenomenon under investigation and for methodological contribution (Ebekozien *et al.*, 2022). A quantitative research method (survey) was used in the first part of the study, and a qualitative research method (interviews) was adopted in the second part of the study.

The respondents for the study were professional quantity surveyors practising in the Nigerian Construction Industry. All respondents were consulting quantity surveyors operating in Lagos State. A list of registered firms and practising quantity surveyors was sourced online from the website of the national and state chapters of the professional body – Nigerian Institute of Quantity Surveyors (NIQS).

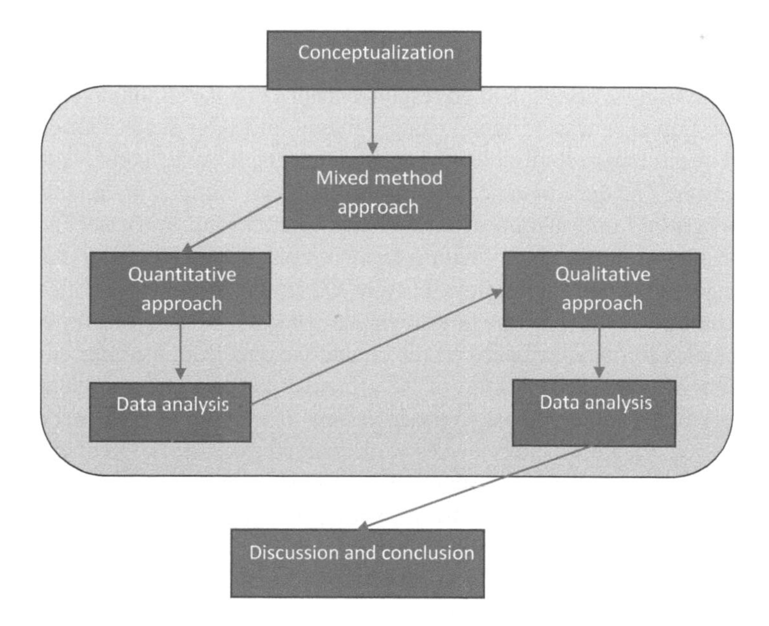

Figure 6.1 Sequential explanatory mixed methods research design adopted for the study

Generally, the respondents were deemed to be qualified based on the following criteria:

1 Consultant quantity surveyors in Lagos, with at least ten years of working experience, and/or registered members of NIQS.
2 Principal and associate partners in QS consulting firms.
3 Principal partners, associate partners, or staff in charge of recruitment in the QS firms.

The structured questionnaires were administered to respondents who met these criteria (1–3) to complete, while the interviews were conducted strictly with those satisfying criteria 2 and 3. The decision to conduct interviews with respondents fulfilling criteria 2 and 3 was to achieve a robust and extensive discussion of the findings from the questionnaire data. Also, these categories of respondents were directly involved in recruiting graduates, hence they possessed first-hand knowledge of the skills requirements, and they determined the needs of the firms for employees and the skills set required, among others. Hence the choice of conducting the interviews with these categories of respondents.

Data Collection Method

The following approaches were adopted to collect data for the study:

- **Quantitative approach:** A well-structured questionnaire was administered to the respondents who were recruited using random sampling. The application of random sampling allowed all members of the population an equal and independent chance of being selected. Also, to attract more participants, respondents were requested to suggest other qualified respondents in a manner similar to snowballing. The questionnaire was designed using a five-point Likert Scale. Out of the total of 150 questionnaires administered to respondents, 33 were considered suitable for analysis. The questionnaire was divided into two sections: in the first section, the background information of respondents was elicited; in the second section, the focus was on questions relating to the employability of QS graduates. The challenges negating the employability of QS graduates were established using the quantitative data from the questionnaire, whereas the solutions for enhancing their employability were based on the qualitative data from the interviews.
- **Qualitative approach:** A total of 22 principal partners and associate partners were contacted to participate in the interview of which nine were accepted and interviewed. Qualified interviewees were sourced from the NIQS databases, and emails were sent to them. Subsequently, those who consented to participate were interviewed. The number of interviewees (9) was considered to be adequate based on existing studies in which it was indicated that qualitative studies can be conducted using samples ranging between 5 and 50 interviewees. Thus, different numbers of interviewees have been adopted in existing studies (Dworkin, 2012; Boddy, 2016; Ebekozien *et al.*, 2022; Hennink & Kaiser, 2022). It was also

shown in these studies that the sample for in-depth interviews in a qualitative study can range from 6 to as many as 22. Thus, the number of in-depth interviews conducted in this study was considered to be adequate. The identity of the interviewees was not revealed in this study

The data analysis was done quantitatively using the Statistical Package for the Social Sciences (SPSS) to determine Mean Scores and a *T*-test. These results are presented in Tables 6.2–6.4. The mean refers to the average score for each of the factors in the data set. The *T*-test was used to compare the differences in the mean scores of the collected data. The mean was used to rank the skills and understand the existing gap. The qualitative data were analysed thematically based on the interviewees' opinions

Results and Discussion

The data analysis and discussion are presented in this section.

Quantitative Data

The quantitative data was classified into two distinct categories: observed competencies and expected competencies. These classifications had been used in existing studies (Jefferies *et al.*, 1992; Alam *et al.*, 2008; RICS, 2015; Vijayalakshmi, 2016). The observed competencies were the competencies detected by employers in QS graduates. This represented the result of the formal structured training received by the graduates in higher education institutions. On the other hand, the expected competencies referred to the list of competencies that employers of staff desire the QS students to possess.

Competences Observed in Recent QS Graduates by Industry Professionals

The data analysis is presented in Table 6.2 that shows 17 competencies ranked by the respondents using a five-point Likert Scale. Trustworthiness was the most observed competency among QS graduates. Other competencies observed included positive attitude, effective communication, strong interpersonal ability, ICT compliance, and information literacy skills.

Based on Table 6.2, it was evident that the top-five competencies observed in QS graduates were behavioural competencies such as trustworthiness, effective communication skills, and information literacy skills. The professional competencies observed were knowledge of contract administration, valuation of work done, estimating, and preparation of preliminary estimates.

It could be inferred that the current curriculum has equipped graduates with the skills highlighted by the employers. A critical analysis of Table 6.2 reveals that communication skills and interpersonal traits dominated the top four most frequently observed competencies in QS graduates.

Table 6.2 Observed competencies in QS graduates

Competency	*MIS*
Trustworthiness	3.78
Positive attitude	3.58
Effective communication	3.53
Strong interpersonal ability	3.52
ICT compliant	3.52
Information literacy skills	3.47
Knowledge of construction technology	3.45
Valuation of work done	3.45
Team player	3.42
Estimating	3.41
Meticulousness	3.36
Creative thinking	3.36
Problem-solving	3.34
Preparation preliminary cost estimate	3.31
Measurement of building services	3.22
Knowledge of construction law	3.22
Cost monitoring	3.13
Managing the tendering process	3.13
Cost advice	3.09
Contract documentation	3.06
Measurement of civil engineering works	3.06
Project management	3.03
Preparing feasibility studies	2.91
Negotiation competence	2.81
Knowledge of QS software	2.76
Knowledge of BIM	2.22

MIS values (where 5, always observed (AO); 4, mostly observed (MO); 3, sometimes observed (SO); 2, rarely observed (RO); 1, never observed (NO)).

Source: Original.

Competencies of Recent QS Graduates Anticipated by Employers

This set refers to the competencies that the industry expects of QS graduates. Respondents were presented with a list of competencies to be ranked using a five-point Likert Scale according to what they considered to be essential competencies possessed by graduates.

Table 6.3 shows the results of the competencies expected of QS graduates. The top four expected competencies observed indicated that more priority was placed on behavioural competencies (trustworthiness, effective communication, meticulousness, and creative thinking). This suggested that industry expectation is for graduates to possess some personal attributes, the ability to recognise when information is needed, and the ability to locate, evaluate, and use information effectively.

Competence Analysis

In this section, the observed and expected competencies are analysed to reveal the gap between the two sets of competencies. A paired sample T-test was conducted

Table 6.3 Competencies expected of QS graduates

Competency	*MIS*
Trustworthiness	4.79
Effective communication	4.55
Meticulousness	4.48
Creative thinking	4.39
Knowledge of construction tech	4.36
Team player	4.36
Positive attitude	4.24
ICT	4.18
Problem-solving	4.18
Strong interpersonal skills	4.18
Information literacy skills	4.10
Estimating	4.03
Measurement of building services	4.00
Contract documentation	3.97
Measurement of civil engineering works	3.88
Valuation of work done	3.88
Managing tender process	3.85
Cost monitoring	3.79
Preparation of preliminary cost estimate	3.73
Knowledge of construction law and regulations	3.70
Cost advice	3.64
Negotiation competence	3.55
Knowledge of QS software	3.52
Feasibility studies preparation	3.44
Project management	3.30
Knowledge of BIM	2.94

MIS values (where 5, highly essential (HE); 4, very essential (VE); 3, essential (E); 2, of little essence (LE); 1, not essential (NE)).

Source: Original.

to test the level of significant difference between the observed and expected competencies.

Pallant (2010) proposed two approaches to the interpretation of a paired sample *T*-test namely:

1 **The probability (*p*) value**: There is a significant difference between the two scores being compared if the value is less than 0.05 (e.g. 0.04, 0.01, 0.001).
2 **Comparing the mean values**: After establishing the significant difference, the next step would be to determine which scores are higher, using the Paired Samples Statistics Table. This gives the Mean Scores for each of the two sets of scores.

It is evident from Table 6.4 that none of the competencies had $p > 0.05$. This indicated a wide gap between the observed and expected competencies acquired by the QS graduates.

Table 6.4 Gap analysis of observed against expected competencies of QS graduates

Competency	*t*	*P*
Effective communication	−5.75	0.000
Creative thinking	−5.51	0.000
Team player	−5.59	0.000
Strong interpersonal skills	−3.75	0.001
Problem-solving	−4.68	0.000
ICT	−3.29	0.002
Meticulousness	−5.40	0.000
Positive attitude	−3.55	0.001
Information literacy skills	−3.18	0.004
Contract documentation	−4.89	0.000
Knowledge of construction tech	−4.75	0.000
Knowledge of construction law and regulation	−2.49	0.018
Managing tender process	−5.04	0.000
Cost monitoring	−4.03	0.000
Cost advice	−3.24	0.003
Preparing feasibility studies	2.82	0.009
Negotiation competence	3.83	0.001
Preparation of preliminary cost estimate	2.13	0.041
Valuation of work done	2.36	0.024
Project management	2.27	0.031
Knowledge of BIM	3.86	0.001
Measurement of civil engineering works	5.89	0.000
Measurement of building services	3.73	0.001

Significance level is $p > 0.05$.

Source: Original.

Enhancing the Employability of QS Graduates

After analysing the quantitative data, a significant difference between the observed and expected competencies was evident. Interviews with principal partners, associate partners, and staff members in charge of recruitment were conducted during the study to discern the steps that could be taken to enhance the employability of QS graduates. Interviewees were asked the following questions:

i Do you employ fresh graduates in your establishment?
ii With your experience, do you plan to employ more? If yes/no, why?
iii Do you consider graduates as fit for the industry? If no, why?
iv What can be done to make graduates fit for the industry by learning institutions?

Theme 1 – Employment of Fresh Graduates

All interviewees agreed to employ fresh graduates in their establishments. However, they indicated that they mostly employed graduates who had prior experience in the form of internships with the firm. Thus, most employers were willing to employ graduates but based on an existing relationship or history with them and their assessment.

Theme 2 – Fitness of Graduates for the Industry

There was consensus among interviewees that QS graduates did not possess the expected competencies. From the transcription of the interviews, the proposed solutions to enhance the employability of QS graduates and make them fit for the industry in the Fourth Industrial Revolution were presented as themes. The following solutions, presented under Theme 3, were suggested by interviewees to bridge the gap between industry and university education.

Theme 3 – Solution to Graduate Fitness for Industry

- Solution 1: Review of the school curriculum to match industry expectations

  This proposed solution was to review the school curriculum to –align it to the industry and, by extension, the curriculum designed to deliver the required competencies to the graduates. According to Interviewee 4, it was observed that graduates lacked business management knowledge and competencies. Consequently, it was proposed that this and many other industry-required skills should be incorporated into the school curriculum. This is important considering the rapid transformation the industry is presently experiencing (Adekunle *et al.*, 2021). Meanwhile, this also included the constant professional development of lecturers to be able to deliver the reviews according to industry dictates.

- Solution 2: The invitation of experienced and practising QS to classrooms to teach students.

  The second solution proposed included the invitation of industry professionals. This was proposed to keep students abreast with prevailing industry standards. Most of the interviewees felt that this would provide students with current exposure to the dynamics and processes of professional practice. According to Interviewee 7, this would give students an understanding of professional perspectives and the dynamics in the field from seasoned and vastly experienced professionals. Interviewee 8 added that it could also serve as a motivation. Examples of courses that industry professionals could teach included ethics and procedures.

- Solution 3: Early, systematic, and periodic exposure of graduates to the practical aspects of QS practice.

  Currently, most QS programmes provide students with industrial experience once at every level of study. Interviewees believed that early exposure is necessary to enable students to gain practical understanding of what they have learnt. Interviewee 2 said, "Schools should enter into a partnership with firms for two weeks rotational internship with firms during academic sessions as part of the curriculum". This would enable students to be aligned with industry practices and processes before graduation. Also, it would enable early introduction to industry trends and development.

Implications for Practice and Research

Studying the employability of QS graduates in a dynamic industry, such as the construction industry which has been impacted by COVID and rapid technological developments, is imperative. The industry is constantly changing together with its graduate skill requirements. A mixed methods research design was adopted in the study to understand the employability challenge and suggest solutions to the identified challenges. The data collected was analysed accordingly. The results revealed that QS graduates were not well-equipped for industry roles. Also, they indicated a difference in perception concerning appropriate competencies between higher tertiary institutions and employers. This study contributes to the on-going discussion about the employability of tertiary institution graduates in the 21st Century. The focus was on the employability of QS graduates from the perspective of employers. This is significant as it provides an industry perspective, which is imperative as a yardstick for establishing the employability of graduates because of the strategic position it occupies.

Furthermore, the research contributes methodologically, as a sequential, explanatory, mixed-methods research design was adopted to achieve the aim. Using this research design ensured a better understanding of the subject matter. The study produced robust and reliable results whilst proffering in-depth solutions to the employability challenges of QS graduates. Also, the method allowed for a systematic inquiry into the research problem and led to possible solutions to the problem. Also, the study flowed from the quantitative approach to the qualitative approach. In other studies, a concurrent MMR approach, or a qualitative MM-driven approach, might be adopted.

Based on the study's results, it was recommended that, first, a periodic assessment of the school curriculum against competence required by the industry should be conducted to provide insights into industry expectations. Also, this would promote the adoption of industry-oriented teaching approaches. This was considered to be essential, especially in the present technology-oriented dispensation with its attendant dynamic state. Continuous professional development of lecturers in line with the industry requirements was also suggested. However, central to these solutions is the provision of the required facilities for training in institutions of higher education.

The implication of this study for graduates is the opportunity to self-assess against employers' expectations. Such assessment would enable them to determine their readiness and suitability for the workplace and industry. Potential limitations of the study include the non-existence of a dedicated human resources department in most QS firms in the study area.

Conclusions

Employability is a critical discourse because of the dynamic nature of the workplace. The chapter contributes to this discussion by including an examination of the employability of QS graduates. A mixed methods approach was adopted to provide a better understanding and a comprehensive solution to the challenge of employability in the construction industry. A wide gap was observed in the study between the expectations of industry and the observed competencies possessed by QS graduates. This suggested that the graduates produced fall below industry expectations and, by extension, do not present a promising outlook for graduate employability.

Consequently, based on the study, an industry-oriented education is recommended, whereby the curriculum is tailored towards meeting dynamic industry requirements. This would prepare the graduates for life after school, rendering them employable. In addition to achieving a balance between professional and behavioural competence, many employers considered this to be a foundation required for a successful professional career. Furthermore, continuous professional development to align the lecturers with prevailing industry trends is imperative as well as the provision of necessary infrastructure to support the efforts of higher learning institutions to promote employability.

References

Adekunle, S.A., John, I. & Aigbavboa, C. 2019. Quantity surveying education for sustainable development. *Industry Perception*, 8: 864–875.

Adekunle, S.A., John, I. & Aigbavboa, C. 2021. Digital transformation in the construction industry: A bibliometric review. *Journal of Engineering, Design and Technology*, 1–29. https://doi.org/10.1108/JEDT-08-2021-0442

Alam, M., Gale, A., Brown, M. & Kidd, C. 2008. The development and delivery of an industry led project management professional development programme: A case study in project management education and success management. *International Journal of Project Management*, 26(3): 223–237.

Aliu, J.O. & Aigbavboa, C. 2020. Employers' perception of employability skills among built-environment graduates. *Journal of Engineering, Design and Technology*, 18(4): 847–864.

Anho, J.E. 2011. An evaluation of the quality and employability of graduates of Nigeria Universities. *African Journal of Social Sciences*, 1: 179–185.

Boddy, C.R. 2016. Sample size for qualitative research. *Qualitative Market Research: An International Journal*, 19(4): 426–432.

Cameron, R. & Miller, P. 2007. Mixed methods research : Phoenix of the paradigm wars. *Proceedings of the 21st ANZAM 2007 Conference: Managing Our Intellectual and Social Capital, Sydney, Australia, 4–7 December 2007*. Promoco Conventions Pty Ltd., Canning Bridge, Western Australia.

Davies, H., Csete, J. & Poon, L.K. 1999. Employer' s Expectations of the Performance of Construction Graduates. *International Journal of Engineering*, 15(3): 191–198.

Dworkin, S.L. 2012. Sample Size policy for qualitative studies using in-depth interviews. *Archaeological Sexual Behaviour*, 41: 1319–1320.

Ebekozien, A., Dominic Duru, O.S. & Dako, O.E. 2022. Maintenance of public hospital buildings in Nigeria – An assessment of current practices and policy options. *Journal of Facilities Management*, 20(1): 120–143.

Hennink, M. & Kaiser, B.N. 2022. Sample sizes for saturation in qualitative research: A systematic review of empirical tests. *Social Science & Medicine*, 292: 114523.

Jefferies, M., Chen, S.E. & Conway, J. 1992. Assessment of professional competence in a construction management problem-based learning setting assessment of professional. *The Australian Journal of Construction Economics and Building*, 2(1): 47–56.

Maarouf, H. 2019. Pragmatism as a supportive paradigm for the mixed research approach: conceptualising the ontological, epistemological, and axiological stances of pragmatism. *International Business Research*, 12(9): 1.

Mansour, B., El & Dean, J. C. 2016. Employability skills as perceived by employers and university faculty in the Fields of Human Resource Development (HRD) for entry level graduate jobs. *Journal of Human Resource and Sustainability Studies*, 4(March): 39–49.

Ornellas, A., Falkner, K. & Stålbrandt, E.E. 2018. Enhancing graduates' employability skills through authentic learning approaches. *Higher Education, Skills and Work-Based Learning*, 9(1): 107–120.Pallant, J. 2010. *SPSS Survival Manual*. 4th edition. McGraw-Hill Companies, New York.

RICS. 2015. *Assessment of Professional Competence Quantity Surveying and Construction*. [Online]. Available at: https://www.rics.org/globalassets/rics-website/media/qualify/pathway-guides/old-pathway-guides/qs-and-construction-chartered-pathway-guide.pdf [Accessed: 18 June 2019].

Srinath, P.,Pearson, J., Zhou, L. & Ekundayo, D, 2013. *RICS Professional Competency Mapping Framework for Programme Appraisal and Benchmarking*. Western Sydney University Open Access Collection, London.

Su, W. & Zhang, M. 2015. An integrative model for measuring graduates' employability skills: A study in China. *Cogent Business & Management*, 2(1): 1060729.

Van Der Heijde, C.M. & Van Der Heijden, B.I.J.M. 2006. A competence-based and multidimensional operationalisation and measurement of employability. *Human Resource Management*, 45(3): 449–476.

Vijayalakshmi, V. 2016. Soft skills: The need of the hour for professional competence: A review on interpersonal skills and intrapersonal skills theories. *International Journal of Applied Engineering Research*. [Online]. Available at: http://www.ripublication.com [Accessed: 18 June 2019].

7 Understanding the Effects of the Built Environment on Autistic Adults

Beth Noble and Nigel Isaacs

Summary

Differences in sensory processing are prevalent among people on the autism spectrum but limited empirical research has considered how they experience the built environment. This explanatory sequential mixed methods study first investigated whether autistic people experienced the built environment differently than neuro-typical people, and whether there were patterns in the effects of different Indoor Environment Quality (IEQ) factors, using an anonymous online survey of both autistic ($n = 83$) and neuro-typical ($n = 134$) participants. Autistic participants reported significantly higher discomfort, distress and avoidance across all environments, together with broad issues related to different IEQ factors. However, in both the Phase 1 survey and the small number of research studies it was not possible to detail specific issues, or in which environments. In Phase 2, a qualitative, participatory photographic approach enabled autistic people to self-select examples of indoor electric lighting that they liked or disliked, send in the photographs and describe why. From this, several areas of interest have been identified, including the design of electric lighting in specific environments, such as medical waiting rooms, for comfort for all users, the identification of physical and emotional reactions that autistic people have to electric lighting, and how people describe lighting that they like and dislike.

Introduction

Everyone is affected by the built environment, and indoor environment quality (IEQ) is linked to comfort, productivity, health, and well-being. Autistic people are widely reported to have differences in how they process sensory information compared with people who are not on the autism spectrum (termed "neuro-typical"), regardless of age or Intelligence Quotient (IQ). However, there is little research about the direct impact of the built environment on adults who are on the autism spectrum.

Working with a diverse group of people requires a novel approach, and an explanatory sequential mixed methods design was used in this research project about the effects of the built environment on autistic people. In this chapter, firstly, the

DOI: 10.1201/9781003204046-7

results of a quantitative survey of a sample of New Zealanders, including autistic and neuro-typical individuals, are reported to evaluate how they respond to various aspects of the built environment. Only selected IEQ factors, such as lighting, acoustics, and thermal comfort, are considered in this chapter. Subsequently, a qualitative, participatory photographic method was used with a small number of autistic adults from New Zealand to identify and understand issues specific to electric lighting, including locations of lighting that cause issues, and the effects of these on autistic people.

Suggestions of solutions to any issues identified were outside the scope of this project, as this study was a first step in understanding effects, both at an initial, broad level and then at a more specific level.

Note: the term "autistic people" is used throughout this chapter as identity-first language is generally preferred by autistic individuals (Kenny *et al.*, 2016). All researchers, including the present authors, must respect the communities of interest, proactively adapting to changing terminology.

Autism and the Built Environment

The autism spectrum is a neuro-developmental condition that affects social and communication skills, as well as patterns of restricted and repetitive behaviour (American Psychiatric Association, 2013). Differences in sensory processing are part of the diagnostic criteria for autism in the most recent revision of the diagnostic manual. These differences are observed in autistic people of all ages and are not dependent on intellectual ability (Leekham *et al.*, 2007; Tavassoli *et al.*, 2014).

Despite there being extensive research into the differences in sensory processing in autistic people, and extensive research into the effects of the indoor environment on the general population, there is limited systematic research that bridges these two areas. Design guidelines have been developed to create "autism-friendly" buildings, but these are often highly specific and are used primarily in specialist schools and residential centres for autistic children who have very high support needs (Humphreys, 2005; Beaver, 2010). Allowances for sensory processing difficulties are specified in all these design guidelines, although to different degrees.

In literature about IEQ, electric lighting has often been assessed as a lower priority than other indoor environmental qualities. However, in most autism-friendly design guidelines, there are some common specifics that appear regularly, including reduction of glare and avoidance of fluorescent lights due to flicker (Humphreys, 2005; Hewitt *et al.*, 2009; Beaver, 2010). In early experimental studies, there was an observed increase in repetitive behaviours in autistic children under fluorescent lights (Colman *et al.*, 1976; Fenton & Penney, 1985). Acoustic consideration is also regularly specified in autism-friendly design guidelines, including the reduction of background noise, consideration of materials, and reduction in reverberation time (Humphreys, 2005; Mostafa, 2008; Hewitt *et al.*, 2009; Beaver, 2010).

The adverse effects of poor IEQ on the general population are well documented. Exposure to poor IEQ can cause both short- and long-term effects across the nervous system, immune system, and endocrine system (Fisk & Rosenfeld, 1997; Bluyssen & Cox, 2002). Poor IEQ also affects productivity negatively, with reductions between 2% and 20% reported, depending on conditions and location (Fisk & Rosenfeld, 1997; Wyon & Wargocki, 2013; Al Horr *et al.*, 2016). However, most available literature about IEQ is based on office workspaces, so these effects may or may not be generalisable to the wider built environment.

Since literature about IEQ is focused primarily on office environments, and existing autism-friendly design is focused primarily on educational settings, the wider built environment used by the whole population day-to-day is largely overlooked, by autistic adults included. Therefore, the aim of the investigation in this mixed methods study was first to identify broad issues that autistic adults face in the built environment, then to focus specifically on exploring further issues with electric lighting.

Methodology

An explanatory sequential mixed methods design was used for this project (Creswell, 2007). The research questions in the project were emergent, in that the research question for Phase 1 was known when the project started, while the research question for Phase 2 emerged from the results of Phase 1 (Plano Clark & Badiee, 2010).

Phase 1 – Quantitative Survey

Research Question 1: Are the adverse effects of the indoor environment exacerbated in autistic people, compared with neuro-typical people?

In Phase 1, participants were recruited through social media, local businesses, word-of-mouth, disability advisory organisations, and local autistic-led self-advocacy organisations. "Autism New Zealand" and "Altogether Autism" both shared the survey throughout their networks, including in adult support groups, local offices, and at a conference. The respondents (self-selected participants) spent 15–20 minutes to complete the anonymous online survey regarding their perception of the indoor environment of their home, workplace, and indoor public spaces. This included questions about discomfort, distress and avoidance.

The survey was based primarily upon the two largest IEQ surveys used worldwide: the Building Use Studies (BUS) Methodology and the Centre for the Built Environment (CBE) Occupant IEQ Survey (Dykes & Baird, 2013). Most items entailed semantic differential ratings or multi-choice questions. Guidelines published by the Australian Co-operative Research Centre for Living with Autism (Autism CRC, 2016) informed the design of the survey, including clear and concise language, consistency, and opportunities for elaboration. Similar to the CBE survey, display logic was used, where questions are displayed to the participant based on previous answers (ASHRAE *et al.*, 2010). For example, to be asked about which types of buildings they avoided, a participant must have answered earlier that they had avoided buildings because of the indoor environment.

A total of 276 responses were collected over two months from July to September 2017. The autistic group (n = 83) was self-identified based on a question in the survey. A further 59 participants who self-identified in the questionnaire as having another condition that affected their sensory processing (e.g. visual impairment, epilepsy, ADHD) or as being related to someone on the autism spectrum, were excluded from the Neuro-typical group (n = 134).

Response data was analysed using the R Studio Statistical Software Programme, which uses the R programming language for statistical computing and graphics. Semantic differential ratings are considered to be discrete ordinal data, as the scale has ordered categories, but the "distance" between the scale points was not uniform, thus the data was not continuous. Many of the distributions were identified as being skewed or non-normally distributed early in the analysis. In particular, the autistic responses often did not form a "bell curve", and were clustered with the majority of responses at one end of the scale.

Therefore, as most measures were ordinal and of skewed distribution, non-parametric analysis methods were used. For a measure of central tendency, the median was used instead of the mean and, for testing between groups, Mann–Whitney U tests were used. Cliff's delta δ was used as a measure of effect size, as it can be used with ordinal data and is a measure of how many times a value from one group is greater than a value from the other group (Cliff, 1993). Cliff's delta produces an effect size between -1 and 1, where 0 indicates no difference between the two groups, while the further from 0, the greater the effect size (Grissom & Kim, 2012). The magnitude of Cliff's delta is related to the effect size magnitudes for Cohen's *d*.

Research Question 2: While results for Phase 1 are reported in detail below, these informed the emergent research question for Phase 2 of the mixed methods approach.

The common themes of the responses of autistic individuals to electric lighting outlined in both the reviewed research studies and from the quantitative first stage of this project were issues of fluorescent lighting, flicker, and brightness. In the quantitative survey, "electric light" as a factor was rated negatively by autistic participants, but the results did not provide specific information about the type of lighting system, layout, or lamp types. It was also not possible to attribute issues regarding glare and electric light to specific building types. Experimental studies also do not specify enough detail to be able to make good design recommendations and are generally limited to educational and institutional environments. Therefore, the emergent research question for Phase 2 was:

What are the characteristics of electric lighting systems that currently meet the needs of autistic people the least, where are they located, and what are their effects?

Phase 2 – Qualitative Photo Diary

A qualitative participatory photographic method was used in Phase 2 (Rose, 2012; Holm, 2014). Participatory photography has precedents of being used with autistic people (Obrusnikova & Cavalier, 2011; Teti *et al.*, 2016) and has been used to evaluate the built environment (Lockett *et al.*, 2005; Tishelman *et al.*, 2016). However, its use for evaluating electric lighting remains novel.

Over the course of two weeks, participants were asked to take photographs of indoor electric lighting and submit them using an online form to create a photographic diary. For each photograph, participants provided an open-ended text description of their thoughts about the lighting photographed, as well as a rating scale of how much they liked/disliked the lights, and answers to questions about location and what type of lighting they thought that it was. After the return period, any necessary clarification and context of the photographs and comments were requested in a follow-up round. This round also provided an exercise to understand technical factors and gave each participant an opportunity to review their individual photographs and descriptive information.

The data from Phase 2 was then broadly coded: first, variables that could be evaluated by the researcher visibly from the photographs and accompanying data forms and second, themes and issues identified from comments.

In the first broad type of coded data, for each of the variables identified, mutually exclusive and exhaustive values were defined (Bell, 2004). These included variables reported directly by participants (e.g. location and type of lamp), and variables evaluated visually by the researcher (e.g. size of space, number of lamps, layout). Numerical content was analysed in using frequency analysis primarily (Bock *et al.*, 2011) but did not include statistical analysis because of the small sample size. The second broad type of coded data included themes and issues identified in the comments on photographs, as well as follow-up responses.

During the initial analysis of the comments on the form for photographs, common themes, and language were identified using thematic analysis (Rose, 2016). This initial data analysis was used to form the questions, clarifications, and exercises in the follow-up. Once follow-up responses had been integrated with the initial data from the photographic diary, further analysis was used to understand how this additional information supplemented the original themes and language.

Photographic diaries were completed by 13 autistic people from September to December 2019, and 11 completed follow-ups in April 2020, after the submission of their photographs. Lockdowns, as a result of COVID-19, were just beginning in April 2020 in New Zealand, so some attrition was not unexpected. The ages of the participants ranged from the 16–17-year-old group to the 55–64 group. Four participants identified as being male (31%), seven as female (54%), and two as gender diverse (15%). Of the 13 participants, 11 reported being formally diagnosed as being on the autism spectrum, while two others were self-diagnosed but reported AQ scores of 45 and 46, well above the indicative threshold for autism of 32 out of 50 (Baron-Cohen *et al.*, 2001).

It is unknown whether the participants in Phase 2 were originally part of the sample in Phase 1 because the Phase 1 survey was anonymous, as is typical in explanatory sequential methods (Creswell, 2007). However, the participants were from the same population – autistic adults from New Zealand.

A total of 136 photographs were submitted by participants (median = 10, min. = 6, max. = 17). Three photographs were excluded from the analysis for not meeting the instruction criteria (i.e. outdoors or duplicates). Therefore, the analysis included 133 photographs of indoor, artificial lighting systems.

Results and Discussion

Phase 1 Results

In the Phase 1 survey, autistic participants reported significantly higher discomfort, distress, and avoidance across all environments. While 90% of participants in the autistic group reported that they had avoided buildings, only 51% of the control group had done so. The probability of an autistic participant having ever avoided buildings was 8.8 times greater (95% CI 4.0, 19.7) than the neuro-typical group ($p < 0.001$).

Participants were asked about discomfort in all building types that they had earlier reported using at least once a month. Eight of the nine building types caused significantly more discomfort for qualifying participants in the autistic group than the neuro-typical group (Table 7.1). The lack of statistically significant difference in how often tertiary education buildings caused discomfort might have been because of the small sample size, as there was a medium effect size between the two groups.

IEQ factors were considered as a cause of discomfort in the home and workplace and as a cause of avoidance in the wider built environment. While 96% of the neuro-typical group completed the workplace section of the survey, only 83% of the autistic group did. This was probably because of the low employment rates documented for autistic people (NAS, 2016). Conversely, 93% of the autistic group completed the avoidance section of the survey compared with 63% of the neuro-typical group. Therefore, while nearly all participants in the autistic group qualified for this question, only two-thirds of the neuro-typical group qualified, and it was expected that these were likely to have been the most sensitive of this group (Table 7.1).

Table 7.1 Comparison of autistic and neuro-typical median frequency of discomfort in public spaces (1 = Never; 7 = Always)

Building type	*Autistic*		*Neuro-typical*		*δ (95% CI)*	*U*	*p*
	n	*M (95% CI)*	*n*	*M (95% CI)*			
Shopping mall	45	6 (5, 7)	94	3 (3, 4)	0.57 (0.38, 0.72)	3325.5	<0.001
Supermarket	75	6 (5, 6)	131	3 (3, 4)	0.53 (0.37, 0.65)	7508.0	<0.001
Medical	41	5 (3, 6)	49	2 (2, 3)	0.51 (0.28, 0.68)	1514.0	<0.001
Other retail	68	4 (4, 5)	125	3 (2, 3)	0.46 (0.30, 0.60)	6207.0	<.001
Restaurant/ bar/café	61	4 (4, 5)	124	3 (3, 3)	0.41 (0.24, 0.56)	5336.0	<0.001
Office	40	5 (4, 6)	90	3 (2, 4)	0.40 (0.19, 0.59)	2531.5	<0.001
Arts	44	3 (2, 4)	58	2 (2, 2)	0.39 (0.17, 0.58)	1779.0	<0.001
Community	30	4 (3, 5)	32	2.5 (2, 3)	0.35 (0.06, 0.58)	648.5	0.016
Tertiary education	22	4 (2, 6)	22	2 (2, 4)	0.33 (0.01, 0.61)	322.5	0.057

Cliff's delta (δ) 0.17 = small, 0.33 = medium, 0.47 = large. CI, confidence interval; M, median.

Source: Original.

Both people and people-noise have the largest magnitudes for both groups in the avoidance section, being the only IEQ factors with a median at the maximum end of the scale for autistic people (Table 7.2). Despite people-noise causing the neuro-typical group very low discomfort in the home, as a cause of avoidance, it was their highest rated factor. In contrast, for the autistic group, people-noise was the IEQ factor that caused the greatest discomfort at home, and was equally the highest with the people factor in both the workplace environment and as a cause of avoidance.

Responses to glare and electric light also had large differences between the autistic and neuro-typical groups across all three IEQ sections (Table 7.2). Glare was rated equally across all three sections by the neuro-typical group. However, for the autistic group, it was associated with increased discomfort in the workplace and was highly rated as a cause of avoidance in the wider built environment.

Temperature was the only IEQ factor that had the same median for both the autistic and control groups across all three sections, with no significant difference between the groups in any environment (Table 7.1). Temperature was the IEQ factor rated highest by the neuro-typical group both at home and at work, which was consistent with thermal factors being widely accepted as a priority in literature about IEQ (Leaman & Bordass, 2007; Wargocki *et al.*, 2012; Al Horr *et al.*, 2016). However, the autistic group rated other factors as causing greater discomfort in both environments. A similar pattern was also evident in the factors of air quality and air movement.

In order to be able to suggest solutions to create more accessible environments for all users, it was important to understand the mechanisms of why the identified issues were a problem. While the results of the Phase 1 survey were useful in an exploratory sense, the conclusions that could be drawn from it were limited. In the

Table 7.2 Comparison of autistic and neuro-typical median discomfort at home and work, and avoidance of buildings

	Home discomfort			*Work discomfort*			*Avoidance*		
	M_A	M_C	U	M_A	M_C	U	M_A	M_C	U
People	2	1	3257.0†	5	2.5	2087.0†	7	5	1854.5†
People-noise	4	1	2326.5†	5	3	2663.5†	7	6	2082.5†
Indoor/other noise	3	1	2786.5†	4	3	2798.0†	5	4	2392.0*
Glare	3	2	3368.9†	4	2	2928.0†	5	2	1859.0†
Electric light	3	2	3755.0†	4	3	3189.0†	5	4	2014.5†
Natural light	2	1	3633.5†	2	2	4247.5	1	1	2474.0*
Temperature	3	3	4996.0*	4	4	4114.0	5	5	2696.5
Air quality	3	1	3529.0†	3	3	3856.0	5	5	2983.5
Air movement	2	2	4209.5*	3	3	4301.0	3	4	2630.0
Smells	3	2	3446.0†	3	2	2934.5†	6	5	3007.5
Privacy	2	1	4022.5†	4	3	3180.5*	5	4	2349.5*
Colour	1	1	4783.5	2	2	3293.0*	4	2	2053.5†

Note. * = $p < 0.05$, † = $p < 0.001$. 95%. M_A = Autistic group median, M_C = Control group median.

Source: Original.

survey, participants were asked about discomfort caused by various IEQ factors in the home and workplace, and how important these were as a cause of avoidance of indoor public spaces. It was not possible to enquire in detail why these effects were reported. In the avoidance section, it was also not possible to attribute specific IEQ factors to specific building types or characteristics.

Phase 2 Results

Data from each photograph and questionnaire (including location, size of space, number of lamps, type, and layout) were evaluated visually and analysed in relation to data based on rating scales. There was a wide variety of building types in the photographs submitted, although all participants submitted at least one photograph taken within their own residence. Overall, nearly half of all the photographs were of lighting that participants disliked, while a quarter was neutral, and a quarter was liked.

Figure 7.1 shows locations where at least 10 photographs were submitted. From these, the most disliked locations were supermarkets, followed by other retail and medical locations, similar to the findings in Phase 1. Most photographs of medical locations were taken in waiting areas – places where it is important for users to be kept calm and relaxed. The high combined responses of dislike for library/community locations were the most unexpected, given the smaller effect found in Phase 1. This category covered many different types of spaces, including library stacks, a theatre, and meeting rooms, so the lighting characteristics seen in the photographs across these locations were also very varied. It was not clear why the proportion of lighting disliked in these locations was high because, unlike supermarket and retail locations which are designed for sales over comfort, library/community locations are presumably spaces where people can relax.

A high proportion of photographs was submitted indicating that the type of light was "Not known" (42%). Participants tended to be reasonably certain about the type of lights in their own homes. They disliked most lights which they reported as being fluorescent (79%) but these were most often tubular shaped in a regular

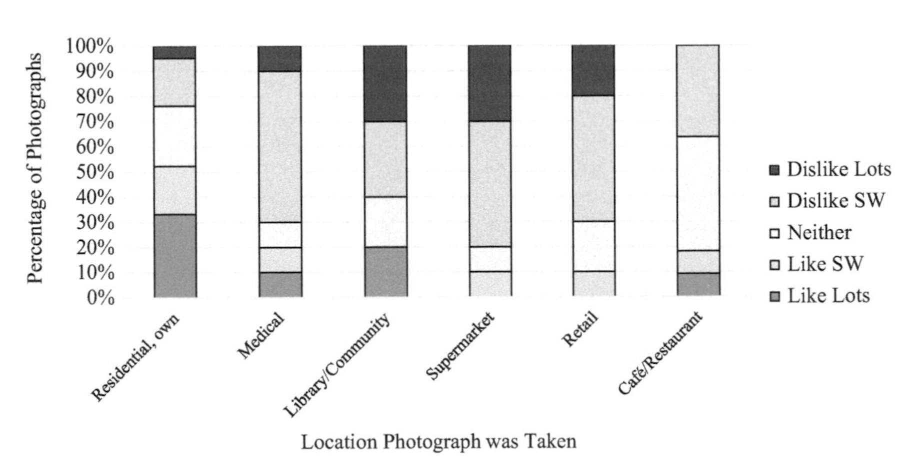

Figure 7.1 Percentage of photographs liked/disliked by location

array and, given advances in lighting technology, it was unclear whether these actually were fluorescent or LED. Generally, participants disliked lighting laid out in a regular array (56%) and, the more lights there were in a space, the higher was the proportion of disliked photographs.

From the comments about the photographs, 18 factors were identified as lighting characteristics that were more objective or measurable and, therefore, potentially able to be included in design guides. These "technical factors" (Table 7.3) were synthesised in the initial photograph coding. Then, in the follow-up, participants were able to select from a list of 18 factors which they felt were a problem and, then again, which they liked, or even a factor that they both liked and felt was a problem. Participants did not have to select every factor.

Table 7.3 shows a graphical and numerical summary and the number of respondents (*n*). Rows shaded in red show that most participants selected the factor as a problem, through to green where most participants who selected the factor liked it. Although some factors were clearly liked, and other factors were clearly identified as problems, there were many factors where responses were mixed, and these differed among participants. Three factors (bright light, cool colour, dim light), had responses that were mixed in that there were some participants who liked the factor and some who found it a problem, but they were heavily skewed towards

Table 7.3 Follow-up exercise with technical factors identified in analysis of photographs

Technical factors	*Like*	*Both*	*Problem*	*Description*	*n*
Flicker	0	0	11	Problem	11
Noise (e.g. buzzing, humming)	0	0	11	Problem	11
Glare	0	0	8	Problem	8
No control over individual lights or dimming	0	0	8	Problem	8
Reflections on other surfaces in the space	1	0	7	Problem	8
Large lights (including multi-tube blocks)	0	0	7	Problem	7
Bright light	1	1	7	Mixed-problem	9
"Cool" (blue) coloured light	2	0	6	Mixed-problem	8
Natural light in a space as well as artificial	7	0	4	Mixed	11
Patterns or shadows	4	0	5	Mixed	9
Lights that hang low	3	0	5	Mixed	8
Different types of lamps in one space	2	1	4	Mixed	7
Lights laid out in a grid	4	0	3	Mixed	7
Several lights in a space	3	0	3	Mixed	6
Lights that are up high	3	0	3	Mixed	6
Dim light	5	1	2	Mixed-like	8
"Warm" (yellow) coloured light	5	0	0	Like	5
Small lights	5	0	0	Like	5

Source: Original.

either being a problem or a liked factor and, as such, these were defined as a "mixed problem" or "mixed like" based on the skew. As noted above, there seemed to be stronger negative responses (problems) than positive ones (likes).

When participants submitted their photographs, they often described their physical or mental/emotional reactions, including positive (6% of the photographs), neutral/moderate (16%), and negative reactions (34%). Not every photograph submitted included a description of a reaction to the lighting in the comments (43% none). However, all participants reported at least one physical or mental/emotional reaction to at least one photograph.

Most reactions noted by participants were negative. All participants mentioned a negative reaction to the lighting in at least one photograph. Negative reactions were varied, although often included reference to headaches, pain, and issues with eyes, including eye strain and "tired" eyes. Most negative reactions were physical, with only a few being specifically mental/emotional. Positive reactions were generally mental/emotional reactions, where participants would describe lighting as "comforting" or "relaxing". Only one photograph included a non-calming, positive reaction, where the participant found the lighting to be "energising".

Six of the 11 participants who completed follow-ups reported that they had previously avoided buildings because of artificial lighting. When questioned further, these buildings included supermarkets, malls, clubs where strobe lighting is used, large retail stores, a work office, and a lecture room at tertiary education. One further participant did note having previously left a job in an office as a result of experiencing headaches from a combination of fluorescent lighting and high noise levels in the environment.

Participants used distinct language to describe their perceptions and opinions of the lighting that they photographed. This use of language involved primarily adjectives – used in sentences such as "lighting…" or "this light". An analysis of descriptive language was not intended in the initial analysis plan but, when the data were collected, it became apparent that the language which people used to describe their perceptions or opinions of lighting was distinct from describing their reactions or identifying technical factors.

Notably, participants would often describe lighting that they liked as having the inverse of qualities that they disliked. The most common example of this was "not too bright" or "not harsh". Participants generally had more varied negative words for lighting than positive words. The common positive words used were "warm" and "soft", as well as the more generic "good" or "great". Interestingly, participants did not regularly use terms such as "appropriate" or "sufficient" to describe lighting that was the correct brightness for them, despite these being the common terms used in IEQ surveys to assess lighting (Dykes & Baird, 2013).

Implications

The explanatory sequential mixed methods approach used in this study provided additional information and qualitative data to help to explain issues identified in the quantitative stage. In research about IEQ, quantitative surveys are commonly used in the assessment of the built environment but are a blunt tool. The addition of in-depth qualitative data made it possible to explore the "what", "where", and "why" issues identified in the survey, particularly as participants were able to self-select what they felt was important to them to submit as photographs and not just answer predetermined questions.

The use of the explanatory sequential mixed methods design used in this study was successful in obtaining results that would not otherwise have been found. The chosen methods were also complementary as, while both provided an opportunity to engage directly with the autistic community, Phase 2 made greater analysis possible, including issues not originally identified as being of interest or concern. The small sample size meant that statistical analysis was always going to be an issue, but the Phase 1 quantitative and Phase 2 qualitative analysis provided a useful combination.

Conclusions

In the study on which this chapter was based, use was made of a quantitative online survey and a qualitative photographic diary. This combination of methods provided valuable new insights into how the built environment affects autistic people, particularly regarding electric lighting.

Phase 1 was an initial step in understanding the effects of various IEQ factors on people on the autism spectrum, considering both the home and workplace environments, as well as the wider built environment. By surveying the perceptions of the built environment of a broad group of people, it was found that people on the autism spectrum do experience greater adverse effects of IEQ. The large group differences observed across many IEQ factors indicated the need for further study into understanding the nature and cause of these effects. In this study, artificial (electric) lighting was selected.

In Phase 2, this work was continued by identifying specifics of the effects of electric lighting on autistic people, including identifying which electric lighting characteristics were least beneficial for them, and where these systems are located. It was found that particular work is needed to improve electric lighting in locations such as medical waiting rooms, where user comfort is a priority for the building occupiers as well as visitors. Also of interest were the types of descriptive language used by people to describe lighting, and the reactions to electric lighting that autistic participants described, both positive and negative.

A question that arises from these results is if indoor environments were designed to consider the needs of autistic people better, how would this affect neuro-typical people? In particular, would neuro-typical people also be more comfortable in an environment where the greater sensory sensitivity of people on the autism spectrum is taken into consideration? If so, designing environments that are accessible to the most sensitive of the population would be likely to improve environments for all users.

References

Al Horr, Y., Arif, M., Kaushik, A., Mazroei, A., Katafygiotou, M. & Elsarrag, E. 2016. Occupant productivity and office indoor environment quality: A review of the literature. *Building and Environment*, 105: 369–389.

American Psychiatric Association. 2013. *Diagnostic and Statistical Manual of Mental Disorders: DSM-5*. 5th edition. American Psychiatric Association. Arlington, VA.

ASHRAE, USGBC, & CISBE. 2010. *Performance Measurement Protocols for Commercial Buildings*. ASHRAE, Atlanta.

Autism CRC. 2016. *Inclusive Research Practice Guides and Checklists.* Version 2. Autism CRC Ltd., Brisbane, Queensland, Australia.

Baron-Cohen, S., Wheelwright, S., Skinner, R., Martin, J. & Clubley, E. 2001. The autism-spectrum quotient (AQ): Evidence from asperger syndrome/high-functioning autism, males and females, scientists and mathematicians. *Journal of Autism and Developmental Disorders*, 31(1): 5–17.

Beaver, C. 2010. Autism-friendly environments. *The Autism File*, 34: 82–85.

Bell, P. 2004. Content analysis of visual images. In: T. Van Leeuwen & C. Jewitt (eds). *The Handbook of Visual Analysis*, pp. 10–34. SAGE Publications Ltd., London.

Bluyssen, P.M. & Cox, C. 2002. Indoor environment quality and upgrading of European office buildings. *Energy and Buildings*, 34(2): 155–162.

Bock, A., Isermann, H. & Knieper, T. 2011. Quantitative content analysis of the visual. In: E. Margolis & L. Pauwels (eds). *The SAGE Handbook of Visual Research Methods*, pp. 265–282. SAGE Publications Ltd., London.

Cliff, N. 1993. Dominance statistics: Ordinal analyses to answer ordinal questions. *Psychological Bulletin*, 114(3): 494–509.

Colman, R.S., Frankel, D.F., Ritvo, E. & Freeman, B.J. 1976. The effects of fluorescent and incandescent illumination upon repetitive behaviors in autistic children. *Journal of Autism and Childhood Schizophrenia*, 6(2): 157–162.

Creswell, J.W. 2007. *Designing and Conducting Mixed Methods Research.* SAGE Publications, Thousand Oaks, CA.

Dykes, C. & Baird, G. 2013. A review of questionnaire-based methods used for assessing and benchmarking indoor environmental quality. *Intelligent Buildings International*, 5(3): 135–149.

Fenton, D.M. & Penney, R. 1985. The effects of fluorescent and incandescent lighting on the repetitive behaviours of autistic and intellectually handicapped children. *Australia and New Zealand Journal of Developmental Disabilities*, 11(3): 137–141.

Fisk, W.J. & Rosenfeld, A.H. 1997. Estimates of improved productivity and health from better indoor environments. *Indoor Air*, 7(3): 158–172.

Grissom, R.J. & Kim, J.J. 2012. *Effect Sizes for Research: Univariate and Multivariate Applications*. 2nd edition. Taylor & Francis Group, New York.

Hewitt, J., Hewiston, C., O'Toole, G. & Haywood, N. 2009. *Sensory Access in Higher Education: Guidance Report 2009*. Equality Challenge Unit, London.

Holm, G. 2014. Photography as a research method. In: P. Leavy (ed). *The Oxford Handbook of Qualitative Research*, pp. 380–402. Oxford University Press, Oxford.

Humphreys, S. 2005. Autism and architecture. [Online]. Available at: http://www.auctores.be/auctores_ bestanden/UDDA%2003102008%20S%20Humphreys.pdf

Kenny, L., Hattersley, C., Molins, B., Buckley, C., Povey, C. & Pellicano, E. 2016. Which terms should be used to describe autism? Perspectives from the UK autism community. *Autism*, 20(4): 442–462.

Leaman, A. & Bordass, B. 2007. Are users more tolerant of 'green' buildings? *Building Research & Information*, 35(6): 662–673.

Leekham, S.R., Nieto, C., Libby, S.J., Wing, L. & Gould, J. 2007. Describing the sensory abnormalities of children and adults with autism. *Journal of Autism & Developmental Disorders*, 37(5): 894–910.

Lockett, D., Willis, A. & Edwards, N. 2005. Through seniors' eyes: An Exploratory qualitative study to identify environmental barriers to and facilitators of walking. *Canadian Journal of Nursing Research Archive*, 37(3): 48–65.

Mostafa, M. 2008. An architecture for autism: Concepts of design intervention for the autistic user. *Archnet-IJAR: International Journal of Architectural Research*, 2(1): 189–211.

NAS. 2016. *The Autism Employment Gap: Too much Information in the Workplace*. The National Autistic Society, London, UK.

Obrusnikova, I. & Cavalier, A. 2011. Perceived barriers and facilitators of participation in after-school physical activity by children with autism spectrum disorders. *Journal of Developmental & Physical Disabilities*, 23(3): 195–211.

Plano Clark, V.L. & Badiee, M. 2010. Research questions in mixed methods research. In: A. Tashakkori & C. Teddlie (eds). *SAGE Handbook of Mixed Methods in Social and Behavioral Research*, pp. 275–304, 2nd edition, Sage, Thousand Oaks, CA.

Rose, G. 2012. *Visual Methodologies: An Introduction to Researching with Visual Materials/ Gillian Rose.* 3rd edition. SAGE, London; Thousand Oaks, CA.

Rose, G. 2016. *Visual Methodologies: An Introduction to Researching with Visual Materials*. 4th edition. SAGE Publications Ltd., London.

Tavassoli, T., Miller, L.J., Schoen, S.A., Nielsen, D.M. & Baron-Cohen, S. 2014. Sensory over-responsivity in adults with autism spectrum conditions. *Autism*, 18(4): 428–432.

Teti, M., Cheak-Zamora, N., Lolli, B. & Maurer-Batjer, A. 2016. Reframing autism: Young adults with autism share their strengths through photo-stories. *Journal of Pediatric Nursing*, 31(6): 619–629.

Tishelman, C., Lindqvist, O., Hajdarevic, S., Rasmussen, B.H. & Goliath, I. 2016. Beyond the visual and verbal: Using participant-produced photographs in research on the surroundings for care at the end-of-life. *Social Science & Medicine*, 168: 120–129.

Wargocki, P., Frontczak, M., Schiavon, S., Goins, J., Arens, E. & Zhang. H. 2012. Satisfaction and self-estimated performance in relation to indoor environmental parameters and building features. *Proceedings of 10th International Conference on Healthy Buildings, Brisbane, Australia*. http://escholarship.org/uc/item/451326fk.

Wyon, D.P. & Wargocki, P. 2013. How indoor environment affects performance. *ASHRAE Journal*, 55(3): 46–52.

8 A Mixed Methods Evaluation of the Social Value of Indigenous Procurement Policies in the Australian Construction Industry

George Denny-Smith, Riza Yosia Sunindijo, Megan Williams, Martin Loosemore, and Leanne Piggott

Summary

Australian Governments seek to use indigenous procurement policies (IPPs) to advance their purchasing power, create social value and address Indigenous Australians' exclusion from the mainstream workforce, including in the construction industry. However, assessing the social value that IPPs create has been limited to economic indicators that have not considered indigenous people's understanding of social value. IPPs have been developed without engaging with indigenous people, further limiting the effectiveness of policy evaluations. In this chapter, the social value created by IPPs is investigated using an Aboriginal evaluation framework called Ngaa-bi-nya. In Ngaa-bi-nya indigenous social value is conceptualised and used to drive relevant mixed methods research of social value. Based on qualitative data collected from 18 group and individual interviews (yarns) to expand and explain the results of an online survey of 150 construction businesses, the findings indicated that IPPs create social value for the indigenous communities for which they are designed. IPPs create social value when indigenous businesses use them strategically to employ indigenous people and to develop meaningful careers in the construction industry. In this study, it is shown how mixed methods evaluation can be used to gain a deeper understanding of the social impacts created by these emerging policies and provide new insights which are valuable to policy-makers, practitioners, and researchers in the field of social procurement.

Introduction

Successive Australian Governments seek to use IPPs to address socio-economic inequities experienced by Indigenous Australians since colonisation. IPPs are a type of "social procurement … to provide opportunities for vulnerable groups which otherwise would not be able to participate in the formal economy on equal footing with others" (Panezi, 2020: 218). Social procurement involves new forms of public governance where the public sector builds strategic partnerships with the private sector to deliver public services and create social value (Raidén *et al.*, 2019). In Australia,

DOI: 10.1201/9781003204046-8

IPPs are used to target social value and "stimulate Indigenous entrepreneurship, business and economic development, providing Indigenous Australians with more opportunities to participate in the economy" (NIAA, 2020: 8). The built environment is critical to these goals because of its size in terms of gross domestic product (GDP) and employment, and pledges to large infrastructure construction committed to by Australian Governments.

Social procurement is also increasingly embedded in public procurement as governments recognise that the issues they are addressing are more inter-twined and "wicked" than anticipated (Barraket, 2021). For example, social procurement policies exist in the UK, Canada, South Africa and New Zealand, in addition to Australian IPPs. Generally, the aim of these policies is to create social value, defined as the importance that people place on the changes they experience in their lives as a result of these policies, taking into account negative impacts (Raidén *et al.*, 2019). However, there are cultural, conceptual, and practical difficulties in defining and assessing social value because it means different things to different stakeholders (Raidén *et al.*, 2019).

A significant challenge faced by social procurement strategies, such as IPPs, is the capacity of organisations to measure the social impacts they target (Barraket, 2021). Assessment of social value is usually left to the stakeholders who design, implement, or fund social procurement policies. This has resulted in a multitude of approaches to assessment. Therefore, assessing social value remains ambiguous in nature. Taylor *et al.* (2021) argued that "realist" impact assessments that use mixed methods research to evaluate how policies affect community outcomes and social well-being are a solution. However, evaluating social value in an IPP context currently adheres to rigid Western economic indicators that exclude the worldviews and experiences of indigenous stakeholders.

In this chapter, the aim is to evaluate IPPs by including community stakeholders to assess social value from their perspectives using Ngaa-bi-nya, an Aboriginal evaluation framework developed by Wiradjuri scholar, Williams (2018), and used successfully in other contexts such as education (Bennie, 2021) and health (Blignault *et al.*, 2021). Drawing on recent research in which the value of mixed methods designs to assess social value is highlighted (Taylor *et al.*, 2021), the participatory approach used to reach a shared agreement between the authors and indigenous stakeholders about research objectives, mixed methods design, methods of data collection, and reporting of social value in an IPP context is explained.

This approach helps policy-makers, practitioners, and researchers to undertake community-centred evaluations about the impacts of the built environment on the communities in which it exists. Researchers can use this approach to develop their own community-based research designs. Practitioners might find it valuable to develop, implement, and monitor their own community engagement or social procurement initiatives. Finally, policy-makers might find it beneficial in designing, implementing, and evaluating new social procurement policies which reflect community values, needs, and priorities better. This is critically important, as public policy in which social outcomes are targeted must acknowledge the socio-economic, institutional, and cultural structures that might influence their success (Fabian *et al.*, 2021).

Social Value Research and Indigenous People

Assessing social value in the built environment tends to be conducted on a once-off basis after a project has been completed. This overlooks the need to identify and include affected stakeholders in the evaluation process to determine outcomes and establish relevant impacts (Nicholls *et al.*, 2012). For Indigenous Australians, this excludes the perspectives and experiences of the workers and communities that IPPs are designed to benefit. Indeed, Denny-Smith *et al.* (2020) argued that poorly designed IPP evaluations might create unintended negative impacts by reinforcing colonial power relationships and recommended that social value assessments in an IPP context should acknowledge and respect the experiences of the indigenous stakeholders they affect.

Evaluating social value in an IPP context has also been recognised as being limited in the recent Indigenous Evaluation Strategy published by the Australian government (Productivity Commission, 2020a: 7), which found that:

- Impacts on Indigenous Australian people are examined in few evaluations of mainstream policies and programmes despite government policy indicating that this should occur;
- Evaluation decisions are made ad hoc, with a limited strategy for their timing, conduct, or involvement of indigenous stakeholders;
- Evaluations have limited quality, including a lack of rigorous methodologies used for evaluations of policies and programmes;
- Evaluation is often an afterthought and not integrated into the policy cycle;
- Indigenous Australians are generally excluded from determining what is evaluated and the design, conduct, and interpretation of evaluations, in which the priorities and perspectives of Indigenous Australians and communities are ignored;
- Evaluations typically are of limited use, in that they tend to be focused on policy or programme-specific questions, with no perspective on the impact that governments are having on the lives of Indigenous Australians.

Collectively, these insights highlight the need for relevant methodologies and methods to evaluate IPPs in which the worldviews and experiences of indigenous stakeholders are recognised, and which inform research and policy that create meaningful outcomes for the businesses and communities affected. This is critically important given the transformative aims of social procurement in colonised lands, such as Australia, where Indigenous Australians have been dispossessed and controlled by colonial governmental systems which dictate how and when economic development should occur (Country *et al.*, 2019). To this end, the Ngaa-bi-nya framework, combined with mixed methods research, can bridge the theoretical and practical gaps in understanding social value.

The value of using mixed methods research is being able to explore the perspectives and experiences of multiple stakeholders by using quantitative and qualitative data in an abductive methodological framework. Some authors have claimed that the effectiveness of mixed methods research is limited by the difficulties of combining quantitative and qualitative methodologies (Salehi & Golafshani, 2010). However, mixed methods research in indigenous-led research projects is effective because of its potential for cultural responsiveness to diversity in perspectives and experiences of participants. As Bardi scholar, Dudgeon, and colleagues recently argued, mixed methods research that combines knowledge and action is effective in building on the strengths and resources of indigenous communities, so that it benefits all research partners (Dudgeon *et al.*, 2020). Most evidence of the value of mixed methods research is found in indigenous-led health research, where mixed methods, including document analysis, surveys, and "yarning" – culturally appropriate interviews – are used to develop holistic models of healthcare services that are culturally informed and appropriate (Bailey *et al.*, 2020). To date, these insights have not been applied in the built environment sector to illustrate the theoretical and practical implications of social value better in relation to IPPs.

Using Ngaa-bi-nya to Inform IPP Evaluations

Recognising the perspectives of indigenous stakeholders on social value requires that researchers prioritise indigenous perspectives in their work. Theoretically, the conceptualisation of social value in this study has been derived from Ngaa-bi-nya. The Ngaa-bi-nya framework promotes Indigenous Australians' ways of being (ontology), knowing (epistemology) and doing (axiology) (Martin & Mirraboopa, 2003). Its value is that it prompts evaluators to consider the historical, policy and social environment of indigenous peoples' lives, thereby informing IPP evaluations that are culturally relevant, effective, and translatable.

The Ngaa-bi-nya framework also encourages the evaluation of Aboriginal programmes, such as IPPs, to be qualitative and quantitative from indigenous peoples' perspectives and captures the social, cultural, and economic influences on the success of programmes such as IPPs. Based on this framework, Denny-Smith *et al.* (2020: 1152) posed the questions in Table 8.1 to evaluate the social value of IPPs. In this chapter, the focus is on the domains of Landscape and Resources as examples to show how a deeper understanding of social value can be achieved in an IPP context. The questions also achieve the following aims: to evaluate IPPs by including community stakeholders to assess social value from their perspectives; and to show how indigenous frameworks might shape mixed methods research with indigenous people in the built environment. In the Landscape domain, the purpose was to explore how IPPs might contribute to socio-economic outcomes experienced by stakeholders. In the Resources domain, the purpose was to identify how indigenous businesses might use IPPs to create social value.

Table 8.1 Ngaa-bi-nya domains interpreted for relevance to evaluating social value from IPPs

Ngaa-bi-nya domain	*Questions to evaluate social value and IPPs*
Landscape	• Has the project promoted self-determining practices of local indigenous people? • Has the project and supply chain improved the socio-economic position of local indigenous people?
Resources	• What Indigenous businesses have been sub-contracted to different work packages on the project? • What employment and training opportunities has the project provided? • What financial outcomes did local indigenous businesses and workers achieve from the project? • How were the skills and experience of local indigenous workers developed during the project?
Ways of working	• How did the project address the social determinants of health and well-being? • How did the project promote cultural identity for workers? • How engaged were the local community during the project and were their concerns addressed?
Learning	• What challenges and setbacks were experienced on the project? • How were they overcome and did this contribute to positive relationships between the contractor and other businesses or the community? • What were the levels of trust, reciprocity and sharing between the contractor and local communities and businesses?

Source: Original.

Research Methodology

To indigenous people, "research" is often associated with colonial histories that extracted and claimed ownership over indigenous knowledge (Smith, 2012). For social value research, this requires researchers to acknowledge that research practices conducted by non-indigenous researchers often ignore indigenous cultures, values, and epistemologies (Foley, 2000) and promote the expertise of researchers at the expense of Indigenous research "subjects" (Productivity Commission, 2020b).

The reflexivity and critical self-reflection of researchers are essential in all research and when working in a diverse, multi-cultural research team. For example, one author of this chapter is Wiradjuri (central New South Wales), while the others are Anglo-Celtic, Indonesian, and Jewish, that is non-indigenous. Cultures influence researcher positionality which involves "cultural assumptions, standpoints and biases" (Martin & Mirraboopa, 2003: 212).

To ensure participatory knowledge-making and actions, and the importance of indigenous ontologies and epistemologies to social value, this study was guided by the principles of community-based participatory research (CBPR). CBPR is a "collaborative process that equitably involves all partners in the research process and recognises the unique strengths that each brings" to a research project (Minkler, 2005: ii3–ii4). In Australian research involving indigenous people, clear ethical research guidelines must be met. These require respecting the research strengths of indigenous people and ensuring that the research is designed, conducted, and reported so that it benefits indigenous people, communities and research partners who have facilitated it (AIATSIS, 2020).

To operationalise the insights above, an explanatory, sequential, mixed methods research design was used in response to the COVID-19 pandemic. Consistent with CBPR, extensive consultation was undertaken with external stakeholders in the public, academic, construction, and other private business sectors to workshop how best to do this research during the pandemic. After ensuring that stakeholders were reminded of the research aim and objectives, almost all advised that the best way to reach the target sample (indigenous businesses that had been engaged on an IPP contract) was to invite potential participants to complete an anonymous online survey about IPPs and social value followed by an invitation to register their interest in being interviewed about social value in an IPP context.

An online survey was developed on the Qualtrics platform, based on Ngaa-bi-nya, the evaluative questions in Table 8.1, and informed by a critical literature review. After workshopping the survey with community and industry stakeholders, minor amendments were made to streamline the survey and improve its accessibility, such as providing clear terms of reference in survey questions. A four-point Likert Scale was used in the survey, which required participants to agree or disagree with statements about the social value created by IPPs and to indicate a position for or against each statement. If participants felt there was no change because of IPPs, they could select "Disagree" in their response. A mean score greater than 2.50 was deemed to indicate support for Ngaa-bi-nya and, therefore, to create social value.

Interview questions based on similar material to the survey were piloted with stakeholders to ensure the same questions were asked to achieve "methodological triangulation", which involves the application of different methods to collect data on the same topic with the aim of demonstrating high validity (Roulston, 2010).

Participants were recruited for the online survey by sending an email to construction businesses on Indigenous Business Directory (IBD), a leading IBD in Australia. A refined search of indigenous construction businesses on IBD was used to recruit participants using non-probabilistic sampling methods. After completing the survey, participants were invited to register for interviews and those who registered were later contacted. Theoretical saturation was achieved after 15 interviews, and three more interviews were completed to validate the findings. The results of this process are presented below.

Results and Discussion

Invitations to participate in this study were sent to 515 indigenous businesses in the construction industry, 150 survey responses were obtained and 22 people indicated their interest to be interviewed. Two groups ($n = 3$ and 2, respectively) and 16 individual interviews were conducted ($N = 18$). Group interviews were used

Table 8.2 Survey sample frame

Category	*Frequency*	*Percent*
Construction or property business		
No	35	23.3
Yes	102	68.0
Missing	13	8.7
Total	150	100.0
Owns or works for an indigenous business		
No	27	18
Yes	108	72.0
Missing	15	10.0
Total	150	100.0
Business owner/manager/employee		
Employee	15	10.0
Business owner/Manager	110	73.3
Missing	25	16.7
Total	150	100.0

Source: Original.

when participants who worked together were available at the same time and the researchers could not establish alternative interview times. Group interviews were a form of "collaborative yarning" that reflects indigenous cultural processes of knowledge-sharing (Bessarab & N'gandu, 2010). The group interviews also demonstrate the flexibility of mixed methods research that enables researchers to adapt to changing conditions in the field. Besides triangulating the results, using qualitative findings to expand on survey findings helps to manage the authors' positionality in research. Presenting the mixed methods results also ensures that data are grounded in the experiences of the participants who contributed to this research.

Of the 150 returned surveys, most respondents (62%) were indigenous. During interviews, some participants, who were non-indigenous business partners, explained that this was because the indigenous partners asked them to participate because of other commitments or because they did not want to participate individually. Table 8.2 shows the survey sample frame.

Following best practice in addressing historical research that reduced indigenous participants according to a number and principles of data sovereignty (Janke, 2021), the ethical documents included a requirement to give participants the informed option to consent to their name being used when writing up the results. Most participants gave their permission, and pseudonyms were used where participants did not clearly indicate their permission to use their names. Table 8.3 shows the qualitative sample frame.

Mixed Methods Capturing Social Value Insights

Social value findings are presented under the Landscape and Resources domains to illustrate the impacts that IPPs have on indigenous businesses and workers.

Table 8.3 Interview sample frame

Name	*Indigenous*	*Indigenous business*	*Industry*	*Role*
Zach*	n	y	Facilities management	Operations manager
Tony*	n	y	Construction	Construction manager
Nathan M	n	y	Construction	Project manager
Nathan A	n	y	Construction and facilities management	Project director
Richard	n	y	Construction	Project manager
Jerri	n	y	Construction	Site engineer
Christopher	n	y	Aboriginal corporation	CEO
Daniel	y	y	Construction and infrastructure	Owner operator
Ash	y	y	Construction and infrastructure	Owner operator
John P	y	y	Construction and infrastructure	Owner
John B	y	y	Construction and infrastructure	Owner
Mike	n	y	Construction and infrastructure	Owner
Tim	y	y	Construction	Owner
Ian	y	y	Architectural and construction management	Owner
Geoff	n	n	Architecture and community planning and engagement	Director
David	y	y	Civil construction	Operations manager
Wayne	y	y	Civil construction	Owner
Sally*	n	y	Recruitment	Owner
Greg*	n	n	Business consulting	Owner
Yarra*	y	y	Media and communications	Owner
LaPa*	y	y	Construction	Owner
Peter*	n	y	Consulting and evaluation	Owner

*Pseudonym used.

Source: Original.

Quantitative data were analysed using descriptive statistics. Descriptive statistics gave initial insight into how the variables corresponded to the analytical social value questions in Table 8.1. Qualitative data were analysed thematically using the social value questions in Table 8.1. This meant that qualitative data could be used to explain the descriptive results of the survey.

Table 8.4 Mean score of landscape responses

Landscape variable	*Mean*
My business's involvement with IPPs helps me be a role model to people in my community.	2.92
My business's involvement with IPPs helps me be a role model to others in my family.	2.91
IPPs have changed my business focus to employing people with values that are similar to my own.	2.84
IPPs have made my company develop better relationships with the local community.	2.81
IPPs have motivated me to be more responsive to my employees' heritage and culture.	2.70
I have employed more indigenous workers because of IPPs.	2.61
IPPs get companies to be involved with local Indigenous communities in a good way.	2.57

Source: Original.

Landscape

Table 8.4 shows the mean score of Landscape responses in the online survey. Most variables showed support for the Landscape domain, meaning that contextual landscape factors influenced the creation of the social value of IPPs. The highest-ranking responses indicated how IPPs helped the respondents to be role models. Interviewees supported this finding. For example, Wayne explained his motivation for capitalising on IPP contracts:

> … now would be the time to stand up and put your hand out…and say I want to do this job, give me some help to do it. And the help is that you will be considered for this work, because … it's got to be spent through indigenous business.

Wayne also explained how IPPs encourage an environment that promotes indigenous leadership, by creating space where people capitalise strategically on opportunities to lead their own businesses and promote socio-economic outcomes for their communities.

Interviewees generally agreed that IPPs create better relationships and employment outcomes in their communities. For example, Tim discussed that they use IPPs strategically to create sustainable employment for indigenous staff: "The apprentices are loving the support and their opportunity…. I'm hoping to have another two in the next couple of years … because … that's our social return, that's our commitment". Tim's reasoning was reinforced by other interviewees who regularly discussed the impact of IPPs on socio-economic outcomes like employment and workplace inclusion. For example, John2 explained how their company leverages IPPs to positively impact the Landscape domain through sustained employment: "… it's [IPP] not only about building the business, but it's the employment".

Table 8.5 Mean score of resources responses

Resources variable	*Mean*
I have good relationships with other companies I've worked with on IPP contracts.	2.95
The income from IPPs means I can provide my employees with more training so they can improve their skills and earn more in the future.	2.49
The income from IPPs means I can pay my employees more, to help them meet their basic needs.	2.49
The income from IPPs means I can pay my employees more so they can do the things they want.	2.30
IPPs create a fun social environment.	2.20

Source: Original.

Resources

Table 8.5 shows the results of responses to Resources survey questions. Most respondents agreed that they had good relationships with contractors they had worked with on IPP contracts. This happened in two ways. The first is that IPPs created space for indigenous contractors to challenge construction-specific biases about their capabilities and leverage that into more work opportunities: "[Government agencies] are using us as a success story. And, because of that, word's getting around. I'm getting phone calls now from people like [Tier one contractors]" (Mike). The business and employment opportunities that IPPs create were critical for businesses to continue to employ and develop indigenous staff for sustainable careers in construction and other socio-economic outcomes. The second way was through stronger connections with the construction industry. Findings indicated that IPPs support the stronger connections that can create social value: "[IPPs are] encouraging big business because I work with companies now … [like Tier one contractor]. You know, they're massive. They do like nearly $2 billion a year. And they're sub-contracting to companies like us" (Ian).

Other business owners capitalised further on industry connections and used IPPs strategically to create opportunities for other businesses too: "… we were definitely working with them [Tier one contractor] to try and encourage … as much indigenous engagement as possible … with [their] targets" (Nathan1). One respondent was more direct in the business's approach to using industry connections to create more opportunities for indigenous businesses. The respondent explained that they used the main contractors' IPP targets as justification for creating more opportunities for indigenous contractors: "*We're working with [Tier one contractor] at the [large infrastructure project] ... and then we said, Why don't we go through the Aboriginal business to help your Aboriginal stream?*"

Some aspects of IPPs that negate the social value they create were also highlighted during interviews. Interviewees discussed several barriers to social value, mainly low engagement from procurement managers, compliance behaviour, and the Black cladding. This was important because they were issues not originally included in the original survey design. It shows the importance of mixed methods research in illustrating variables that might otherwise have been overlooked.

Participants had experienced low levels of engagement from procurement managers in relation to IPPs. Interviewees explained that procurement managers had little awareness of indigenous business. As one survey respondent wrote, "So few procurement officers [know] what an IPP is, what it mandates they do and what parameters they have to ensure indigenous inclusion in their procurement". An interviewee explained that "The [procurement managers] wouldn't know who's in the local area within 100 kilometres" (Wayne). The low engagement experienced by participants inhibited social value because procurement managers would select the non-indigenous contractors they knew specifically. This limits social value because indigenous contractors are prevented from winning sustainable work.

Another concern for participants was the practice of "Black cladding", which occurs when a larger non-indigenous contractor forms a joint venture company with an indigenous shareholder to take advantage of government procurement opportunities. While the indigenous partner/s own/s at least 50% of the company to qualify as an indigenous business, they retain little to no control over the business operations and strategy, while the business employs little to no indigenous people (Mundine, 2016). "The goodness of the IPP is being diluted by many black clad businesses and, therefore, margins are tightening and the ability to give back is also lessening" (survey respondent). Black cladding significantly limits the ability of IPPs to create social value because it takes money away from businesses that genuinely want to create social value: "They've [black clad businesses] got no indigenous staff. They're not hiring.... I'm not saying I'm perfect, but ... I'm definitely chasing and hiring indigenous staff" (Tim).

Implications for Practice and Research

In this study, it has been shown that explanatory sequential mixed methods research design can improve the rigour and validity of community-based research in the built environment and develop the nascent social value scholarship in this field. Mixed methods research contributed to collaborative, community-based research, the purpose of which was to promote indigenous experiences and standpoints. This addresses the historical exclusion of indigenous peoples from evaluations that affect them (Productivity Commission, 2020b). Ethical research with indigenous stakeholders is also based on what works best from the perspective of the indigenous stakeholders who guide research processes (Dudgeon *et al.*, 2020). Therefore, the pragmatic and flexible nature of mixed methods research is well-suited to research about indigenous social value and research about the built environment with diverse stakeholders generally. When used in partnership with community stakeholders at the "cultural interface" of Indigenous and Western Science (Nakata, 2007), mixed methods research makes space for the co-design of research. This, in turn, might involve the perspectives and experiences of indigenous stakeholders more equitably while also adhering to Western academic requirements for reliability and validity in ethical research.

By using an explanatory sequential mixed methods research design in this study, it was found that IPPs have a positive impact on the Landscape and Resources domains for indigenous businesses and employees. Indigenous construction businesses use IPPs strategically to employ more Indigenous staff and promote construction careers. These findings challenge research in health where it was found that indigenous health-workers were often restricted to low-paid roles and unstructured career pathways (Bailey *et al.*, 2020).

Mixed methods research also provided an important, yet overlooked, dimension to the performance of IPPs in the Australian Construction Industry. For example, Australian Governments espouse the performance of IPPs against their contract targets as evidence that the policies are effective in creating the business and employment opportunities to which the policies aspire (NIAA, 2021). However, it was found that indigenous contractors have lost work to black-clad businesses, thus challenging the notion that IPPs are a universal solution to addressing the historical exclusion of Indigenous Australians from business and work.

Findings also demonstrated empirically the value of Aboriginal frameworks in evaluating indigenous programmes quantitatively and qualitatively. By using Ngaa-bi-nya, with mixed methods research to report the perspectives of indigenous stakeholders in relation to social value and IPPs, it was found that indigenous contractors used IPPs strategically to create economic and social values in their workplaces and communities. They capitalised on IPP opportunities to invest in sustainable construction careers that create indigenous leaders for others to follow. The culturally sensitive mixed methods research, governed by Ngaa-bi-nya, and involving the indigenous stakeholders in its development, revealed an important finding because it proved the benefits of culturally informed workplaces in the construction context. Until now, this has been limited to research in health (Bailey *et al.*, 2020) and the Australian Public Service (Bargallie, 2020).

For evaluation and social value research and practice, mixed methods research offers further, important, new insights into the value of using frameworks, such as Ngaa-bi-nya, to perform culturally appropriate research. This could overcome the continued racial discrimination against Indigenous Australians in government policies and their evaluations (Thurber *et al.*, 2021). Further, effective social value assessment espouses the importance of involving stakeholders and mixed methodologies to evaluate social value accurately, culturally appropriate frameworks are also needed to report on appropriate social value outcomes.

Conclusions

The purpose of this chapter was to contribute to mixed methods research and practice by evaluating IPPs to assess social value from the perspectives of community stakeholders. While demonstrating a culturally informed evaluation that assesses social value from the perspective of indigenous stakeholders, a combination of quantitative and qualitative data using the Ngaa-bi-nya framework helped to evaluate IPPs for social value. Findings showed how mixed methods can give a deeper understanding of social phenomena by explaining quantitative results qualitatively through the experiences of participants.

The value of this mixed methods research lies in the insight it has given to the impacts created by new IPPs in Australia. IPPs create social value when indigenous businesses use them strategically to lead the way for sustained employment in the industry and invest in the skills and professional development of indigenous staff. However, the benefits of IPPs to the Landscape and Resources domains of Ngaa-bi-nya are being offset by the adverse effects of black cladding, which challenges the existing narrative that IPPs are a universal success. Indigenous contractors can flourish when they are unencumbered by the limitations on their business activities created by black cladding.

Methodologically, this research has been important for establishing guidance on using mixed methods to understand the impacts of social procurement on affected stakeholders as the field matures. Mixed methods research helps researchers to be aware of their positionalities in research with marginalised populations. This is important for anyone working in multi-cultural research teams, cognisant of power differentials and the rights of indigenous peoples to self-determine research, policies, and programmes that affect their lives.

References

AIATSIS. 2020. *AIATSIS Code of Ethics for Aboriginal and Torres Strait Islander Research.* Australia Institute for Aboriginal and Torres Strait Islander Research, Canberra.

Bailey, J., Blignault, I. & Carriage, C. 2020. *We are Working for Our People: Growing and Strengthening the Aboriginal and Torres Strait Islander Health Workforce.* Lowitja Institute, Melbourne.

Bargallie, D. 2020. *Unmasking the Racial Contract: Indigenous Voices on Racism in the Australian Public Service.* NewSouth Publishing, Sydney.

Barraket, J. 2021. *The State of Social Procurement in Australia and New Zealand 2021.* Centre for Social Impact, Melbourne.

Bennie, A. 2021. "We were made to feel comfortable and … safe": Co-creating, delivering, and evaluating coach education and health promotion workshops with Aboriginal Australian peoples. *Annals of Leisure Research*, 24(1): 168–188.

Bessarab, D. & Ng'andu, B. 2010. Yarning about yarning as a legitimate method in indigenous research. *International Journal of Critical Indigenous Studies*, 3(1): 37–50. https://doi.org/10.5204/ijcis.v3i1.57

Blignault, I., Norsa, L., Blackburn, R., Bloomfield, G., Beetson, K., Jalaludin, B. & Jones, N. 2021. "You can't work with my people if you don't know how to": Enhancing transfer of care from hospital to primary care for Aboriginal Australians with chronic disease. *International Journal of Environmental Research and Public Health*, 18(14): 7233. https://doi.org/10.3390/ijerph18147233

Country, T., Lee, E. & Eversole, R. 2019. Rethinking the regions: Indigenous peoples and regional development, *Regional Studies*, 53(11): 1509–1519.

Denny-Smith, G., Williams, M. & Loosemore, M. 2020. Assessing the impact of social procurement policies for Indigenous people. *Construction Management and Economics,* 38(12): 1139–1157.

Dudgeon, P., Bray, A. & Darlaston-Jones, D. 2020. *Aboriginal Participatory Action Research: An Indigenous Research Methodology Strengthening Decolonisation and Social and Emotional Wellbeing*. The Lowitja Institute, Melbourne.
Fabian, M., Agarwala, M. & Alexandrova, A. 2021. *Wellbeing Public Policy Needs More Theory*. Bennett Institute for Public Policy, University of Cambridge, Cambridge.
Foley, D. 2000. Indigenous research, differing value systems. *The Australian Journal of Indigenous Education*, 28(1): 17–30.
Janke, T. 2021. *True Tracks: Respecting Indigenous Knowledge and Culture*. NewSouth Publishing, Sydney.
Martin, K. & Mirraboopa, B. 2003. Ways of knowing, being and doing: A theoretical framework and methods for indigenous and indigenist re-search. *Journal of Australian Studies*, 27(76): 203–214.
Minkler, M. 2005. Community-based research partnerships: Challenges and opportunities. *Journal of Urban Health*, 82(SUPPL. 2): ii3–ii12.
Mundine, N.W. 2016. Making a real difference: Does the corporate sector need to lift its game? *Policy: A Journal of Public Policy and Ideas*, 32(1): 11–14.NIAA. 2020. *Indigenous Procurement Policy*. National Indigenous Australians Agency, Canberra.
NIAA. 2021. *Indigenous Procurement Policy*. National Indigenous Australians Agency, Canberra.
Nicholls, J., Lawlor, E. & Neitzert, E. 2012. *A Guide to Social Return on Investment*. The SROI Network, Liverpool.
Panezi, M. 2020. The complex landscape of indigenous procurement. In: Borrows, J. (ed). *Indigenous Peoples and International Trade*, pp. 217–247. Cambridge University Press, Cambridge.
Productivity Commission. 2020a. *Indigenous Evaluation Strategy*. Productivity Commission, Canberra.
Productivity Commission. 2020b. *Indigenous Evaluation Strategy*. Canberra: Productivity Commission, Canberra.
Raidén, A., Loosemore, M. & King, A. 2019. *Social Value in Construction*. Routledge, London.
Roulston, K. 2010. *Reflective Interviewing: A Guide to Theory and Practice*. Sage, London.
Salehi, K. & Golafshani, N. 2010. Using mixed methods in research studies: An opportunity with its challenges. *International Journal of Multiple Research Approaches*, 4: 186–191.
Smith, L.T. 2012. *Decolonizing Methodologies: Research and Indigenous Peoples*. 2nd edition. Zed, London.
Taylor, C.N., Mackay, M. & Perkins, H.C. 2021. Social impact assessment and (realist) evaluation: Meeting of the methods. *Impact Assessment and Project Appraisal*, 39(6): 450–462.
Thurber, K., Colonna, E. & Jones, R. 2021. Prevalence of everyday discrimination and relation with wellbeing among Aboriginal and Torres Strait Islander adults in Australia. *International Journal of Environmental Research and Public Health*, 18(12): 6577.
Williams, M. 2018. Ngaa-bi-nya Aboriginal and Torres Strait Islander program evaluation framework. *Evaluation Journal of Australasia*, 18(1): 6–20.

9 A Methodological Application to Construction Economics Research for Theory Refinement and Extension

Mohd Azrai Azman and Carol K. H. Hon

Summary

Econometric models to establish causal relationships that describe empirical phenomena using economic data are often used in research about construction economics. However, econometric modelling alone cannot unveil the underlying mechanisms and contextual information from the built environment's perspective. Therefore, to fill this gap, it is demonstrated in this chapter that mixed methods sequential explanatory (MMSE) research design can provide valuable explanatory elements in detail. MMSE design has been employed in a limited amount of research about the built environment to provide a deeper understanding that could not be explained by econometric models alone. How MMSE can support the results from econometric models with in-depth insights from experts is discussed in the chapter. MMSE design can improve the results based on in-depth qualitative details received from the experts who are part of the decision-making process. Therefore, recommendations with reference to real-world practice can be made. In this chapter, the first author's doctoral research is used as a case study to illustrate the application of MMSE design. The study started with dominant econometric models in the first stage, and then directed content analysis (DCA) from follow-up interviews with experts were conducted in the second stage. The chapter serves as a guide for researchers to apply MMSE design in research about construction economics and the built environment with the prospect of theory refinement and extension.

Introduction

Mixed methods research (MMR) has gained traction in research about built environment over the years. Research about the built environment is often interdisciplinary, drawing knowledge from natural science, social science, engineering, and humanities. Therefore, researchers in the built environment tend to adopt both quantitative and qualitative research methods to tackle their research questions, making MMR the most appropriate approach (Day & Gunderson, 2018). Despite this, the adoption of MMR in the construction economics of research about the built environment has been limited.

DOI: 10.1201/9781003204046-9

In this chapter, the first author's doctoral research on diversification and productivity performance in large construction firms in Malaysia is used as a case study. Large Malaysian construction firms are most likely to be diversified in different products and overseas markets. Nevertheless, whether product diversification (PD) and internationalisation can improve their productivity performance in the long run needs to be considered. Besides, in developing economies, rapid changes in institutional environments and the variety of different ownership concentrations can contribute to different outcomes.

However, quantitative results cannot be used to unravel the underlying mechanisms on how the diversification strategy affects the firms' productivity performance. In this case, the application of MMSE research design would help in theoretical refinement and extension based on perspectives from managers who made the decision to diversify into related construction business, unrelated to construction and the firms' internationalisation in overseas markets.

In this chapter, the process of developing an explanatory model is demonstrated that can provide a meaningful explanation for the causes of the effect from theoretical and practical perspectives. Subsequently, managers and policymakers could make better-informed decisions based on practical recommendations from the research.

The Impact of Diversification on Productivity Performance in Large Construction Firms

In the current literature about construction, conflicting results are revealed about whether different diversification strategies have a positive impact on firm productivity. In South Korea, PD improves a firm's value but increases its risks of insolvency (Lee *et al.*, 2016). In Spain and Portugal, only highly diversified firms benefit from PD (Horta *et al.*, 2016). In the United Kingdom, focused and moderately diversified firms outperform those that are heavily diversified, in terms of profitability (Ibrahim & Kaka, 2007). Major firms in the United States (US) and internationally, with a high level of PD, produce greater profits and growth than more focused and broadly diversified firms (Choi & Russell, 2005; Yee & Cheah, 2006). In contrast, high-performing construction firms in the US are more likely to diversify, but the more they diversify, the less growth the company experiences (Kim & Reinschmidt, 2011). A high degree of PD increases the asset growth of international construction firms while reducing profit and increasing risks (Sung *et al.*, 2017).

The results in internationalisation are also inconsistent; they show a significant and positive relationship between the firms in Spain but have no impact on the firms in Portugal (Horta *et al.*, 2016). Jang *et al.* (2019) found a positive impact of internationalisation on firm profits but a negative impact on revenue and market share. In addition, PD has a positive effect on internationalisation-revenue and internationalisation-market share. Higher internationalisation decreases the firm's liquidity and increases its debts, thus posing high risks for large construction firms in South Korea (Lee *et al.*, 2016). The top global construction firms are those that expanded in the overseas market with PD and internationalisation (Han *et al.*, 2010; Ye *et al.*, 2017; Sung *et al.*, 2017). Internationalisation negatively moderates PD-corporate social responsibility (CSR) performance (Ma *et al.*, 2016). Previous studies show mixed results in both impact of PD and internationalisation on firm performance. Also, there is a limited result on the impacts of PD and internationalisation on productivity performance. Using productivity performance could unravel the theoretical benefits of mix-and-scale economies that are still currently debated (Hashai, 2015).

The inconsistency in previous results and the lack of focus on the institutional context could offer a new perspective to unveil the effectiveness of firm diversification (Santarelli & Tran, 2016). In developing economies, changes in institutional environments are more pronounced, which can moderate the impact of strategy on productivity performance. Therefore, the institutional context could be a suitable framework for explaining the effectiveness of diversification in developing economies (Holmes *et al.*, 2018).

Different ownerships could be compatible but also in conflict with institutional environments because of conflicting objectives, motivations and incentives. It is argued that, even though government-controlled firms have a formal commercial purpose of maximising productivity, informally, they have to cater to objectives and incentive structures that are politically, socially or nationally inclined, which could devour their commercial prospect. For family-controlled firms, even though they might resemble ordinary privately-controlled firms, the incentive structure could also be different, such as embedded patrimonialism in the firm, which could affect the strategy's effectiveness and render regulatory institutions ineffective.

An in-depth explanation can be drawn from managers' experience as part of a firm's decision-making process to explain the significant effect (Chittoor *et al.*, 2015). For example, the underlying mechanisms of ownership concentrations that would affect the relationship between diversification and productivity performance must be uncovered using a second-stage, qualitative method. This provides more explanation, which can be ascertained from empirical evidence (Simon, 2009). These underlying factors might not be explained by the first-stage, quantitative method. From the literature review, these underlying factors are uncovered in limited studies, and most research is focused on hypothesis testing. Therefore, in this study, MMSE design can provide a more lucid explanation which might bring about theory refinement and extension.

Research Methodology

MMR design requires a combination of at least one quantitative and one qualitative method. The design of MMR could also be one of the important aspects that might determine a researcher's philosophical worldview in research (Schoonenboom & Johnson, 2017). In MMSE design, it is also possible to mix different methods and paradigms that suit the research questions. However, the emphasis in the MMSE design adopted in this case study was on reductionism and theory in advance (Creswell & Clark, 2018). In most cases, quantitative approaches, such as econometric methods for hypothesis testing, have been adopted in studies about construction economics. This deductive approach ensures irrefutable results from the scrutiny of probability-based statistics (Wilkinson, 2013). Nevertheless, insightful explanations would be needed to understand the underlying mechanisms that might not be well understood in the construction industry, where the nature of business and informal contexts might need further explanations (Dainty, 2008). Therefore, sequentially, MMSE design provides a deeper understanding and explanation that could not have been explained by quantitative analysis alone (Ritchie *et al.*, 2014). The design starts with a quantitative (QUAN) method in the first stage and, sequentially, the follow-up qualitative (QUAL) method (Creswell & Clark, 2018) (see Figure 9.1).

The follow-up qualitative method, using interviews, was conducted with individuals who could explain the relationships between the variables (Schoonenboom & Johnson, 2017). More explanation was ascertained from empirical evidence, especially enquiry in the background, leading to the outcomes (Simon, 2009). For example, how the mechanism of variable X would impact variable Y and how the mechanism of Variable Z moderates the impact of variable X on variable Y (see Figure 9.2). In this case, the interviews were used to validate the result and further extend and refine existing theory.

Case Study

In this chapter, a case study of previous research carried out by Azman (2020) to illustrate the application of MMSE design was used. In the first phase, the design starts with the Generalised Method of Moments (GMM) for modelling the impact of diversification strategies (X-variables) on the productivity of 86 large construction firms (Y-variables), and the moderating effects of formal institutional environments (Z_1-variables) and ownership concentrations (Z_2-variables) over a 14-year period (2003–2016). Diversification strategies are dependent on other variables which introduce endogeneity bias. With the presence of endogeneity, covariance between residual (e_{it}) and predictors are not equal to zero.

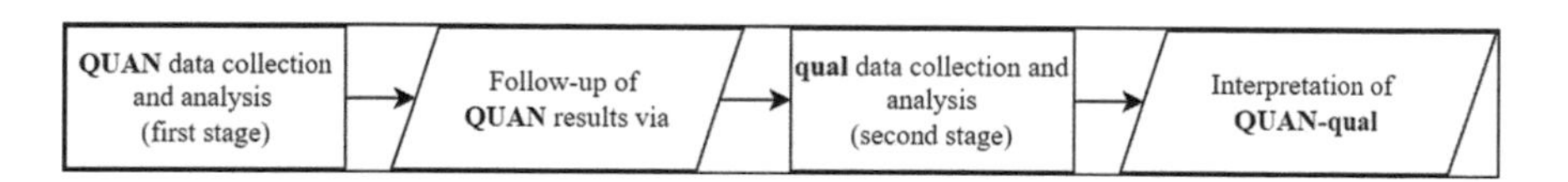

Figure 9.1 Mixed method sequential explanatory research design

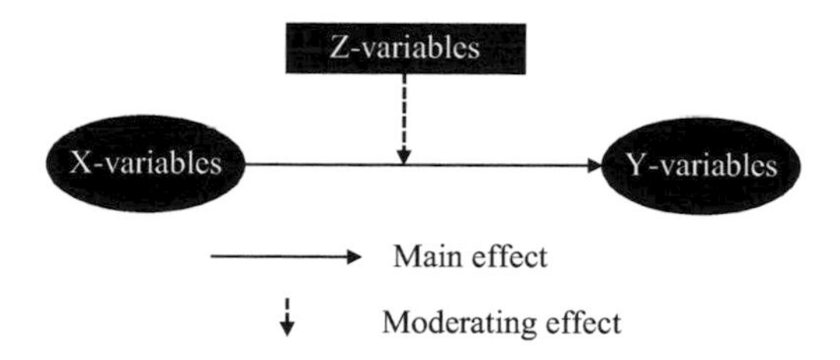

Figure 9.2 Example of a theoretical framework

Table 9.1 The variables of the model

Variable	*Explanation*
X-variables	RD, UD, internationalisation
Y-variables	Firm Productivity (Total Factor Productivity)
Z_1-moderating variables	LCA, LRC, LFD and LPCF
Z_2-moderating variables	FCC and GLCC

Note: RD, related diversification (measure of construction sector-related product diversification); UD, unrelated diversification (measure of non-construction sector product diversification); internationalisation, overseas market expansion; LCA, level of capital availability; LRC, level of regulatory control; LFD, level of foreign debt; LPCF, level of price control freedom; FCC, family-controlled concentration; GLCC, government-linked companies' concentration.

Source: Original.

Therefore, Equation (9.1) indicates violation of the assumptions of ordinary least squares, which requires the application of GMM estimator:

$$\operatorname{cov}\left(e_{it}, x_{kit}\right) \neq 0 \tag{9.1}$$

An inappropriate estimator can mislead the second-stage interviews because the questions posed might arise from false positive (Type I error) or negative (Type II error) quantitative results. GMM uses internal instruments of lagged variables within the datasets. GMM estimator minimises $Q_N(\theta)$, the weighted sum of squares of covariances between instruments and residuals so that $E\left[f(x_i, \theta)\right] = 0$ as defined in Equation (9.2):

$$Q_N(\theta) = \frac{1}{N}\sum_{i=1}^{N} f(x_i, \theta)' W_N \frac{1}{N}\sum_{i=1}^{N} f(x_i, \theta) \tag{9.2}$$

where

x_i includes equations containing unknown parameters and variables (including lag variables as instruments)

W is the optimal weighting matrix.

The x_i in Equation (9.2) can be re-arranged to obtain the estimate $\hat{\beta}$ parameters.

In the second phase, based on purposive sampling, six senior managers were interviewed to validate and explain the first-phase GMM results in depth. Refer to Azman *et al.* (2019, 2021, 2022) for further readings in Table 9.1.

Research Approach

Figure 9.3 describes a summary of the practical processes in applying the MMSE approach based on the case study of this chapter.

For MMSE research, the QUAN results served as the basis for follow-up interview questions in the second phase (Creswell & Clark, 2018). In this case study, significant results were used in follow-up questions. Therefore, the interviews were

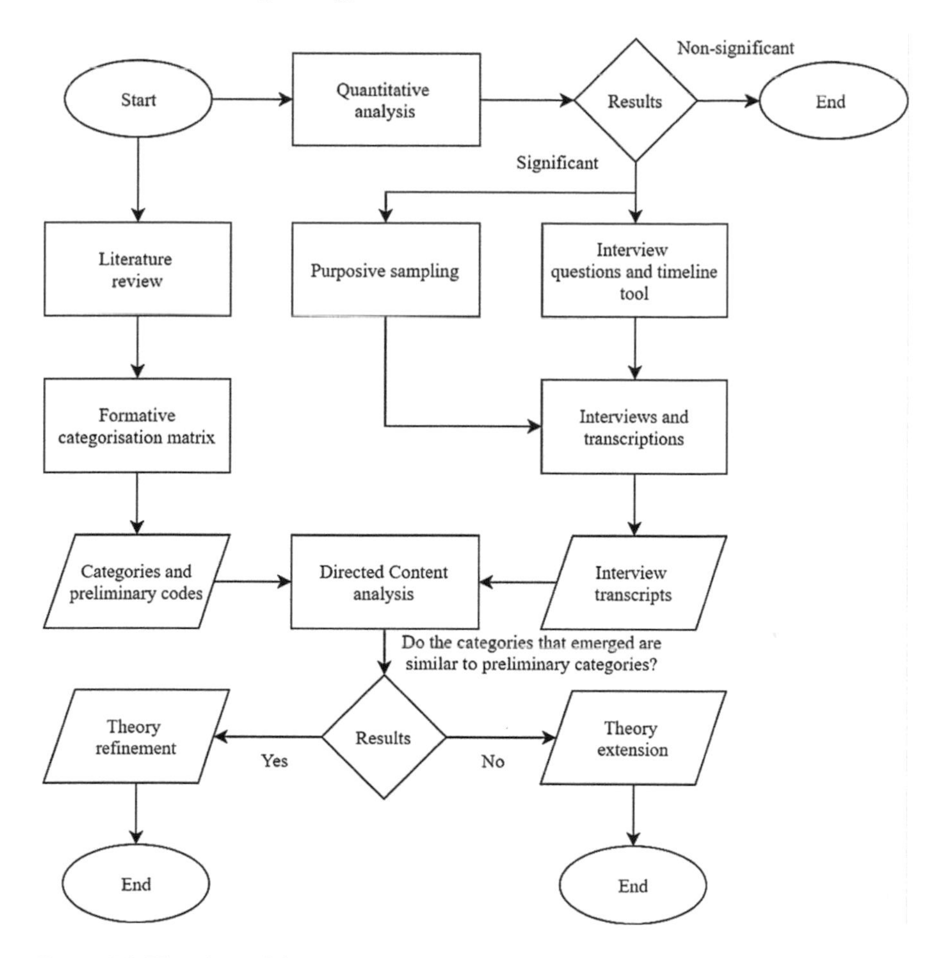

Figure 9.3 Flowchart of the MMSE process

focused on variables, which were statistically significant when $p < 0.10$. According to Creswell (2013), follow-up questions can also be based on standard deviation, extreme cases, group differences and surprising results. It is noted that the quantitative method used by researchers should be objective and less susceptible to model mis-specification and endogeneity.

Interviewing participants about past events could also add several types of errors. Over-reporting (false positive) can be caused by both commission and forward telescoping. In this case, the interviewees tend to overstate their answers based on something they do not experience or report events that happened recently (Kjellsson *et al.*, 2014). Under-reporting (false negative) is caused by omission and backwards telescoping. The interviewees tend to understate their answers even though they have experienced an event that happened before.

In this case study, the timeline technique was used, which is an aided recall technique, to help interviewees recall the related event in the past (Du Plooy, 2009). The interviewees were provided with a graphical timeframe representing history to help them recall their long-term memory (Van der Vaart, 2004). This encouraged the interviewees to recount the timing of the events via visual or mental means.

Figure 9.4 shows the graphical timeframe which displayed the changes in institutional dimensions in different formal, institutional environments from 2003 to 2016. This became the time dimension, thematic axes or landmark events (Glasner & Van der Vaart, 2009). In this study, the terms of Prime Ministers were proposed as the public landmark events: Fifth Prime Minister (2003–2009) and Sixth Prime Minister (2009–2016). This was important because more liberal approaches in finance and business from 2010 and onwards considerably affected the economic and regulatory institutions (post-transition).

A sample of interviewees was selected purposively (Creswell, 2013). If the sample of interviewees is carefully selected based on predetermined selection categories, they can provide an in-depth explanation of the quantitative results (Creswell & Clark, 2018). In this study, senior management of large construction firms which have high diversification thresholds, were selected for interview. Firms with a diversification threshold above the 75th percentile were deemed to be high. Based on the questions prepared, eight categories of firms were selected for follow-up interviews because they could fulfil the high threshold in Table 9.2.

All interviewees were purposively invited based on the above eight categories. Six interviewees agreed to participate in face-to-face interviews. This number of interviewees was considered to be sufficient because they covered the eight categories described above. Table 9.2 shows that interviewees were experienced construction professionals and they had, on average more than 20 years of experience. Also, they worked at the selected categories of firms that were used in quantitative analysis (refer to Table 9.3).

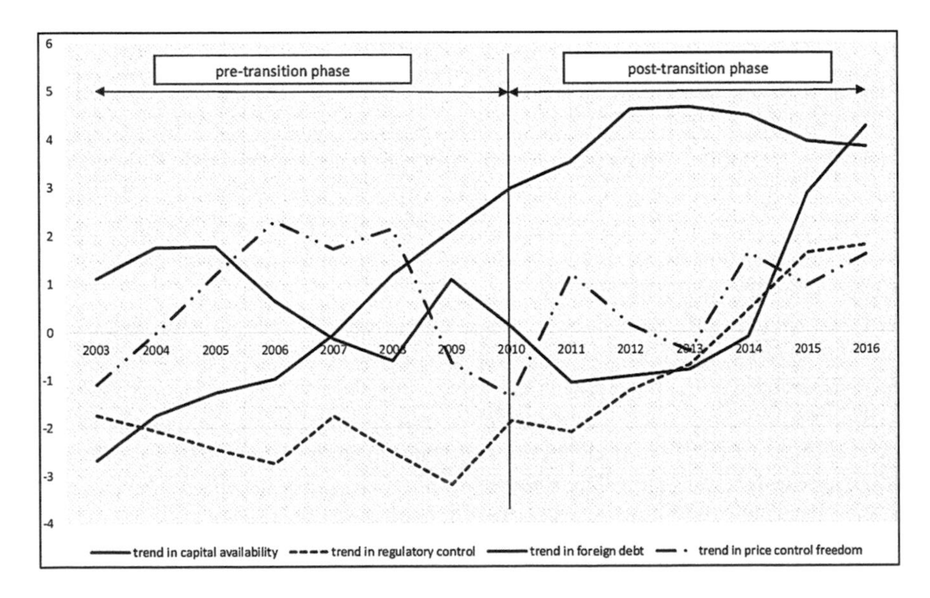

Figure 9.4 Predicted values of institutional dimensions

Table 9.2 Interviewees' profiles

Interviewee	*Position*	*Years of experience*
R1	General Manager (Investment)	14 years
R2	General Manager (Operation)	28 years
R3	General Manager (Business Development)	20 years
R4	Chief Financial Officer	26 years
R5	General Manager (Business Development)	24 years
R6	General Manager (Business Development)	18 years

Source: Original.

Table 9.3 Interviewee profile and categories of firm diversification

	C1	*C2*	*C3*	*C4*	*C5*	*C6*	*C7*	*C8*
R1	✓	✓	✓	✓				✓
R2	✓	✓	✓		✓	✓	✓	
R3	✓	✓		✓				
R4			✓				✓	
R5	✓		✓	✓				✓
R6	✓	✓	✓		✓	✓	✓	

Note: R is the interviewee profile, C is the category of firm diversification (can be more than one) and ü if yes.

Source: Original.

The interviews were conducted and recorded on a confidential basis. The interviews were transcribed by the researchers. The transcriptions were included in interview reports, and the reports were sent to the interviewees for verification (Saunders *et al.*, 2015). Based on significant QUAN results, the following were some of the interview questions:

1 Why does related diversification (RD) have a positive impact on firm productivity from 2010 to 2016?
2 Why did unrelated diversification (UD) hurt firm productivity from 2003 to 2016?
3 Why did internationalisation hurt firm productivity from 2003 to 2009?
4 Why does high level of regulatory control (LRC) in the market affect the positive impact of RD on firm productivity?
5 Why does low level of capital availability (LCA) affect the positive impact of internationalisation on firm productivity?
6 Why does high family-controlled concentration (FCC) affect the positive impact of RD on firm productivity?
7 Why does high FCC affect the negative impact of UD on firm productivity?
8 Why does high government-linked companies' concentration (GLCC) affect the negative impact of RD on firm productivity?
9 Why does high FCC affect the positive impact of internationalisation on firm productivity?
10 Why does high GLCC affect the negative impact of internationalisation on firm productivity?

Content Analysis from the Follow-Up Interview

There are three types of content analysis, namely conventional (inductive), summative and directed (deductive) approaches. A DCA is more appropriate to analyse the interview data because the research aims to validate or extend the theory employed (Hsieh and Shannon, 2005). This study adopted the DCA proposed by Assarroudi *et al.* (2018) and developed formative categorisation matrices for the interview data accordingly.

Formative Categorisation Matrix

The derivation of categories was classified deductively according to the theory, which was related to the research questions and the QUAN results of the study. In the case study, the main explanatory and moderating variables were the main categories. Each of the main categories had preliminary codes identified from the literature, which could be the causes of the effects. According to the literature, it was emphasised that the technology (expertise) [code: MD1] to oversee international operations affected firm productivity. This technology included knowledge and managerial competence within the manager's control. Nevertheless, during the decision-making process, there were environmental variables [code: MD2] that were usually outside the control of managers (Table 9.4). With regard to the moderating effect of the type of ownership concentration, two sub-categories of informal norms can emerge that might influence the relationship between internationalisation and firm productivity, that is substitutive, and complementary or competing and accommodating effects (Table 9.5).

Results and Discussion

Theory Refinement

Based on the interview findings, the formative categorisation matrix from Table 9.5 was refined further. Therefore, Figure 9.5 shows the relationship hierarchy developed to establish the link between main categories and different sub-categories that provide validity and refine the explanation of the quantitative results. Anchor samples or meaning units of interviewees' explanations were provided to illustrate the interviewees' narratives on the categories studied.

Table 9.4 Example of formative categorisation matrix for economies of internationalisation

Main category	*Subcategory*	*Preliminary codes*
Economies of internationalisation	Technology	• Knowledge and competency embodied by the firm (managers) to manage international operations (MD1)
	Environment	• Variables-uncontrolled by managers (firms) (MD2)

Source: Original.

Table 9.5 Example of formative categorisation matrix for ownership concentration that is embodied by informal norms

Main category	*Sub-category*	*Preliminary codes*
Ownership concentration	Informal norms – Substitutive/ complementary (positive effect)	• Efficiently allocate/co-ordinate internal resources to close-knit firms to resolve external market failures (INS1). • Enable checks and balances over managerial opportunism through a more vocal and unique incentive structure (non-monetary benefits of control) (INS2).
	Competing/ accommodating (negative effect)	• Carried out business/product based on national and social goals rather than profitability/productivity objective function (INC1). • Invest/buy-in a business/product owned by others/close-knit for private benefits rather than profitability/productivity objective function (INC2).

Source: Original.

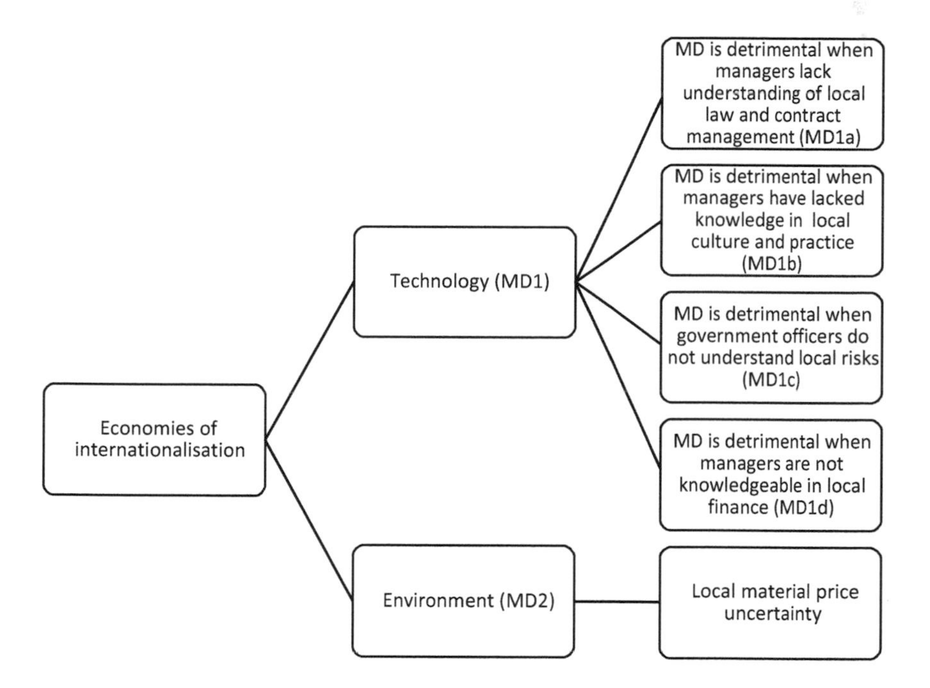

Figure 9.5 Relationship hierarchy of economies of internationalism

Based on quantitative results, the impact of internationalisation on firm productivity was negative during the pre-transition phase. All interviewees agreed with this result. Internationalisation is detrimental **when managers lack an understanding of local law and contract management (MD1a),** thus affecting firm productivity (refer to Figure 9.5). The explanations provided by the interviewees added refinement based on the context of construction firms in Malaysia. Interviewee 1 described how the land acquisition for a public infrastructure project in one of the countries in South Asia was a problem that jeopardised the whole work progress:

> The land acquisition [in this South Asian country] was not clear. In Malaysia, land acquisition is very clear. There is a compulsory acquisition, so the landowner only has the right to challenge the value of the land. You do not have the right to deny the acquisition.

There was a certain condition that only local sources of supplies were allowed in South Asia. Interviewee 2 described that "Sometimes we cannot employ [our own] Malaysians because [this South Asian country's] law requires that only the locals are to be employed".

Interviewee 4 explained that there was a control issue concerning the firm venture in one Middle East country and, at the same time, it also involved compliance costs that continued to be incurred:

> We had a problem because our firm could not hold the major shares of the joint venture; it was difficult for us. You know, we had to pay some sponsorship fees during the project period. Even though the project was completed, the sponsorship fees kept on incurring. We managed 100% of the project, but the other 51% shares were owned by our local partner from the Middle East, and this company was actually a sleeping partner. We paid the sponsorship fee and did all the work ourselves.

Most large construction firms acquired overseas construction projects through government-to-government (G2G) arrangements, awarded through an economic agreement between Malaysia and foreign countries. Normally, government officials make the initial arrangement between the two countries. Therefore, **internationalisation is detrimental when government officers do not understand local risks (MD1c)**. According to Interviewee 1:

> Many companies do not dare to invest in their own – many of them acquired it through the G2G arrangement. Usually, our government also does not make a thorough study of risks and returns before signing the agreement. When our company recorded a loss once, we are now having second thoughts about acquiring projects from overseas.

Theory Extension

Figure 9.6 illustrates a new category that can be extended from the theory (highlighted in grey colour) and emerged from the analysis because it was found to be different from the generic categories established (refer to Table 9.5). Based on quantitative results, the impact of FCC on RD-firm productivity was positive. All interviewees agreed that FCC incentivises RD because **the family is well-informed in the construction-related industry, thus the decision can be made based on trust and collective action (INC3)**.

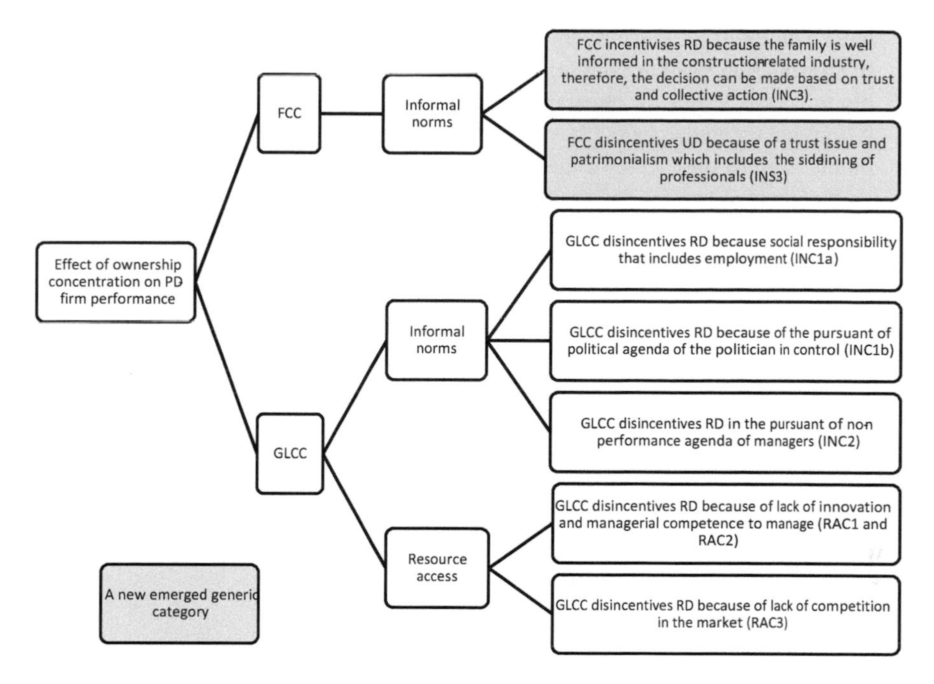

Figure 9.6 Relationship hierarchy of ownership concentration on PD-firm productivity

Interviewee 2 explained that a quick decision was made based on collective action:

> In a family firm, decision making is easier to make especially if it is related to the construction industry, because they have the experience and knowledge in the industry for so long, and the board will make a quick decision as the board members also come from the same family.

Interviewee 6 described that family-controlled firms, embedded with construction-related knowledge, would help them in making the decision:

> You need a business-driven family that drives the business, and they know the industry very well because they started in this business in the early 80s. They know where to get the jobs; they understand what the industry is like. It is good when they are focusing on engineering and construction, but when they do other business, for example, after being so long in the contracting business, now they diversified in the plantation business, it would be a different thing. That is why you see adverse outcomes. What you know in construction might not be relevant to the other industry.

In this case, Interviewee 6 added that, with regard to UD, the family firm must rely on professionals and conflict of trust is a problem:

> Yes, when it comes to unrelated business like a plantation business, you have to rely on the external party like a plantation advisor; the plantation advisor might

> have a different idea on how things should work. For example, people involved in construction want to do things at the most minimal possible cost, and possibly they want to buy the cheapest in the market, for instance fertiliser; however, it does not mean that using the cheapest fertiliser will give proper growth to the plants. Before they were in the contracting business, the family was never in a plantation. When relying on professionals, there could be a conflict of trust.

Based on quantitative results, the impact of FCC on UD-TFP is negative. All interviewees agreed that **FCC disincentivises UD because of a trust issue and patrimonialism, which includes the side-lining of professionals (INS3).** Similar to INC3, a new category emerged from the analysis that could extend the existing theory. Interviewee 2 added that UD could fail because of the family-manager ignorance:

If we venture into other fields, which are not the core business, we have to do a detailed study. Many of the failures occur because the owners think they know everything, for example when a company ventures into a plantation, which is not its core business, yet asks relatives or friends who are also inexperienced to run it, it will invite failures because they don't have the expertise.

Implications for Practice and Research

Based on second-stage results, RD was associated with the technology that managers and firms could easily access and manage, which was within the scope of the construction business. In contrast, UD was associated with technology that was little known to most managers. Thus, RD would improve firm productivity. For internationalisation, it could hurt firm productivity because the strategy required in-depth knowledge of the local context that most managers did not understand well.

There was a possibility that family ownership was more market-oriented, which was also helped by the collective action of family members, such as sharing inputs based on trust. Therefore, high FCC had a positive impact on RD and internationalisation-firm productivity. However, in terms of UD, it was difficult for family firms because they seemed to have little idea of how to run UD more productively. The problem of trust emerged when inputs were needed from professionals outside their family members. Thus, the negative impact of UD on firm productivity could be expected.

The commercial objective of government-linked companies (GLC) was similar to that of private firms. However, in the case of GLCC, for RD or internationalisation, the strategy could be affected by personal, political, and non-competitive arrangements within a firm, affecting the effectiveness of the strategy. Thus, high GLCC hurts RD and the productivity of an internationalised firm. However, because of the close connection with the government, the firms could set their prices to escape losses.

As illustrated in the case study, the MMSE approach is useful in validating the quantitative results. The explanatory elements bring refinement to the current theory, including economic theory regarding the diversification of construction firms in the context of Malaysian construction firms. In addition, MMSE research enabled the extension of theory based on new categories that emerged as it gave new meaning compared with the generic categories developed from the formative categorisation matrix earlier. Therefore, MMSE approach helped to make a clear contribution to knowledge systematically in this case.

Conclusions

The case study was used to illustrate how the MMSE approach could be applied to a construction economics study where in-depth explanations based on qualitative results provide validation and contribute to theoretical refinement and extension. Also, the explanatory elements are useful for the government, policymakers, and other stakeholders to understand the investigated causal relationship better from the contextual perspective of the real world.

References

Assarroudi, A., Heshmati Nabavi, F., Armat, M.R., Ebadi, A. & Vaismoradi, M. 2018. Directed qualitative content analysis: The description and elaboration of its underpinning methods and data analysis process. *Journal of Research in Nursing,* 23(1): 42–55.

Azman, M.A. 2020. *Diversification, Institutions, and Productivity Performance of Large Construction Firms in Malaysia.* Queensland University of Technology, Brisbane.

Azman, M.A., Hon, C.K., Skitmore, M., Lee, B.L. & Xia, B. 2019. A Meta-frontier method of decomposing long-term construction productivity components and technological gaps at the firm level: Evidence from Malaysia. *Construction Management and Economics*, 37(2): 72–88.

Azman, M.A., Hon, C.K., Xia, B., Lee, B.L. & Skitmore, M. 2021. Product diversification and large construction firm productivity: The effect of institutional environments in Malaysia *Engineering, Construction and Architectural Management*, 28(4): 994–1013.

Azman, M.A., Hon, C.K., Xia, B., Lee, B.L. & Skitmore, M. 2022. Internationalization and productivity of construction firms in a developing country: The effect of institutional environment and ownership on Malaysian construction firms. *Journal of Management in Engineering*, 38(1): 04021081.

Chittoor, R., Kale, P. & Puranam, P. 2015. Business groups in developing capital markets: Towards a complementarity perspective. *Strategic Management Journal*, 36(9): 1277–1296.

Choi, J. & Russell, J.S. 2005. Long-term entropy and profitability change of United States public construction firms. *Journal of Management in Engineering*, 21(1): 17–26.

Creswell, J.W. 2013. *Research Design: Qualitative, Quantitative, and Mixed Methods Approaches.* Sage Publications, Los Angeles, CA.

Creswell, J.W. & Clark, V.L.P. 2018. *Designing and Conducting Mixed Methods Research.* Sage, Los Angeles, CA.

Dainty, A. 2008. Methodological pluralism in construction management research. In: Knight, A. & Ruddock, L. (eds). *Advanced Research Methods in the Built Environment*, pp. 1–12. Blackwell Publishing Ltd., Oxford.

Day, J.K. & Gunderson, D. 2018. Mixed methods in built environment research. In: Sulbaran, T. (ed). *54th ASC Annual International Conference Proceedings.* Minneapolis.

Du Plooy, G.M. 2009. *Communication Research: Techniques, Methods and Applications.* Juta and Company Ltd., Cape Town.

Glasner, T. & Van der Vaart, W. 2009. Applications of calendar instruments in social surveys: A review. *Quality and Quantity*, 43(3): 333–349.

Han, S.H., Kim, D.Y., Jang, H.S. and Choi, S. 2010. Strategies for contractors to sustain growth in the global construction market. *Habitat International*, 34(1): 1–10.

Hashai, N. 2015. Within-industry diversification and firm performance - An S-shaped hypothesis. *Strategic Management Journal*, 36(9): 1378–1400.
Holmes, R.M., Hoskisson, R.E., Kim, H., Wan, W.P. & Holcomb, T.R. 2018. International strategy and business groups: A review and future research agenda. *Journal of World Business*, 53(2): 134–150.
Horta, I.M., Kapelko, M., Oude Lansink, A. & Camanho, A.S. 2016. The impact of internationalization and diversification on construction industry performance. *International Journal of Strategic Property Management*, 20(2): 172–183.
Hsieh, H.-F. & Shannon, S.E. 2005. Three approaches to qualitative content analysis. *Qualitative Health Research*, 15(9): 1277–1288.
Ibrahim, Y.M. & Kaka, A.P. 2007. The impact of diversification on the performance of UK construction firms. *Journal of Financial Management of Property and Construction*, 12(2): 73–86.
Jang, Y., Kwon, N., Ahn, Y., Lee, H.-S. & Park, M. 2019. International diversification and performance of construction companies: Moderating effect of regional, product, and Industry diversifications. *Journal of Management in Engineering*, 35(5): 04019015.
Kim, H.-J. & Reinschmidt, K.F. 2011. Diversification by the largest US contractors. *Canadian Journal of Civil Engineering*, 38(7): 800–810.
Kjellsson, G., Clarke, P. & Gerdtham, U.-G. 2014. Forgetting to remember or remembering to forget: A study of the recall period length in health care survey questions. *Journal of health Economics*, 35: 34–46.
Lee, S., Tae, S., Yoo, S. & Shin, S. 2016. Impact of business portfolio diversification on construction company insolvency in Korea. *Journal of Management in Engineering*, 32(3): 1–9.
Ma, H., Zeng, S., Shen, G.Q., Lin, H. & Chen, H. 2016. International diversification and corporate social responsibility: An empirical study of Chinese contractors. *Management Decision*, 54(3): 750–774.
Ritchie, J., Lewis, J., Nicholls, C.M. & Ormston, R. 2014. *Qualitative Research Practice: A Guide for Social Science Students and Researchers*. Sage, London.
Santarelli, E. & Tran, H.T. 2016. Diversification strategies and firm performance in Vietnam. *Economics of Transition*, 24(1): 31–68.
Saunders, M.N.K., Lewis, P. & Thornhill, A. 2015. *Research Methods for Business Students*. Pearson Education Limited, Essex.
Schoonenboom, J. & Johnson, R.B. 2017. How to construct a mixed methods research design. *KZfSS Kölner Zeitschrift für Soziologie und Sozialpsychologie*, 69(2): 107–131.
Simon, H.A. 2009. *An Empirically-Based Microeconomics*. Cambridge Books, Cambridge.
Sung, Y.-K., Lee, J., Yi, J.-S. & Son, J. 2017. Establishment of growth strategies for international construction firms by exploring diversification-related determinants and their effects. *Journal of Management in Engineering*, 33(5): 04017018.
Van der Vaart, W. 2004. The time-line as a device to enhance recall in standardized research interviews: A split ballot study. *Journal of Official Statistics*, 20(2): 301.
Wilkinson, M. 2013. Testing the null hypothesis: the forgotten legacy of Karl Popper? *Journal of Sports Sciences*, 31(9): 919–920.
Ye, M., Lu, W., Ye, K. & Flanagan, R. 2017. How do top construction companies diversify in the international construction market? *Proceedings of the 20th International Symposium on Advancement of Construction Management and Real Estate*. Springer.
Yee, C.Y. & Cheah, C.Y. 2006. Fundamental analysis of profitability of large engineering and construction firms. *Journal of Management in Engineering*, 22(4): 203–210.

10 Effects of Physical Infrastructure on Practical Performance of Graduates in Architectural Technology in Southeast Nigeria

An Explanatory Sequential Mixed Method Research Design

Udochukwu Marcel-Okafor

Summary

The development of architectural technology programmes in relation to technological innovations is reflected in intricate linkages between technology and architecture. Yet, a central concern in education for architect remains the relationship between what is taught in schools and the skills required for effective practice. Consequently, the purpose of this chapter was to outline the practicality of explanatory sequential mixed methods research design in assessing the effects of physical infrastructure on the practical performance of graduates in architectural technology with a view to obtaining comprehensive results necessary to proffer robust solutions. Three objectives were pursued in relation to the impact on the practical performance of graduates in architectural technology; the abilities of architectural technologists to accomplish the stipulated learning outcomes of the HND programme; the state of physical infrastructure; and the relevance of professional courses and entrepreneurial skills taught. Surveys using structured questionnaires were carried out initially to provide a quantitative description of developments in three research populations. In the second phase, a qualitative multiple case study approach was used to explicate further how distinct variables in the objectives related to the practical performance of graduates. The choice of this research design yielded inclusive results for the study, and it is expected to offer useful insights for novice researchers into the adoption and application of the design.

Introduction

Architectural technology programmes in Nigeria are offered exclusively in the Polytechnics and Colleges of Technology (National Board for Technical Education [NBTE], 2016). The aim of polytechnic education is to provide practice-based learning, including the public service of producing a skilled labour force. The NBTE has the authority to review, grant or deny accreditation to programmes in

DOI: 10.1201/9781003204046-10

architectural technology offered by technical institutions in Nigeria. A qualified architectural technologist is trained to assist architects, thereby filling the manpower requirement at the middle level in the chain of architectural practice (National Board for Technical Education, 2016).

The programme in most advanced nations has evolved over the years from meeting the requirements of the earlier role of draughtsman to becoming a contemporary discipline. The advancement of technology-driven designs in the building industry gave rise to architectural technology as a professional discipline in which the responsibilities of the architectural technologist are emphasised in the United Kingdom (Emmitt, 2002). Developments in computer-aided design and draughting (CADD), such as building information modelling (BIM), have emphasised the roles of architectural technologists across the entire lifespan of buildings from the outset of the proposed drawings. The emergence and eminence of BIM are closely associated with the specialisations of architectural technologists, which are anchored in design and project management techniques inter-connected with the lifespan of buildings through a complex bond with technology (Emmitt, 2009). As a profession, architectural technology supersedes other professions in determining the nexus between technology and architecture (Armstrong *et al.*, 2013). The capacity of this discipline to control design strategically in relation to construction processes during the lifespan of buildings ensures that buildings remain cost-effective, well-organised, and functional (Armstrong & Allwinkle, 2017).

There have been concerted efforts to align the practice of architectural technology in Nigeria with the practice of Information and Communication Technology (ICT in contemporary economies by adopting technologically advanced techniques with the aim of providing solutions to the existing challenges within the built environment. It has also been revealed in studies that because architectural practice in Nigeria has been largely computerised, only graduates who are proficient in CADD are relevant and employable (Oladapo, 2017). However, the attempts to conform to international best practices have not been matched with the infrastructural facilities needed in the departments to impart the necessary knowledge and skills (Olotuah *et al.*, 2016; Opoko & Oluwatayo, 2015).

The purpose of this chapter was to illustrate the application of an explanatory sequential mixed methods research design in assessing the effects of physical infrastructural facilities in schools on the practical performance of graduates in architectural technology from Polytechnics in Southeast Nigeria. A methodical approach is provided in the chapter on how to use this type of mixed methods research design to gain an inclusive understanding of a phenomenon.

During the course of this study, answers were sought to the following research questions:

RQ1: To what extent have learning outcomes/objectives of the programme influenced the abilities of the graduates working in architectural firms in Southeast Nigeria?

RQ2: To what extent have the physical infrastructural facilities in the schools affected the abilities of the graduates working in architectural firms in Southeast Nigeria?

RQ3: Are the professional courses and entrepreneurial skills taught relevant to the dynamics of practice in Southeast Nigeria?

To guide the study in the QUAN phase, the following hypotheses were proposed and the variables written in italics were subsequently analysed.

$H_0$1 There is no significant relationship between the *practical performance of technologists* and *proficiency in CADD software.*

$H_0$2 There is no significant relationship between the *number of available functional computer workstations* and *practical performance of technologists.*

Architectural Technology Programme and Practical Performance

In assessing performance in architectural professional practice, the criteria used to measure the extent of services rendered are applied mostly (Oyedele & Tham, 2007) although the particular set of criteria totally explicates their performance. However, the objectives/learning outcomes of programmes for graduates of architectural technology in polytechnics, which represent the criteria for measuring their practical performance, were adopted in this study. The six objectives below are focused on drawing skills and constitute the variables in the study:

1 Assisting in the design and preparation of spatial relationship, circulation, and area diagrams.
2 Developing and preparing details of design drawings.
3 Reading and interpreting technical drawings and specifications, and incorporating such details in the architectural working drawings.
4 Preparing spatial programmes and working drawings for architectural projects.
5 Detailing simple building components in wood, plastic, metal, and concrete.
6 Making presentation drawings of architectural works and drawing detailed perspective with necessary rendering and requisite colour scheme.

Graduates, academics, and employers in the architectural profession agree that the most relevant and highly sought-after skills are adroit computing abilities, creativity, analysis and synthesis of ideas and forms, pragmatic application of knowledge on paper and on the ground, and critical and coherent perspicacity (Maina & Salihu, 2016). Doyle (2019) identified six top skills required for successful architectural practice, of which three are stated below for the purpose of this study:

1 Ability to design buildings with a focus on functionality and stability: This demands a sound knowledge of advanced mathematics and physics to determine the strengths and properties of building materials.
2 Ability to design aesthetically appealing buildings: Man's desire for aesthetics is fluid and, thus, creativity is critical. Broad knowledge of arts and the history of styles is important to actualise this skill.
3 Ability to work with computers: The use of software applications for architectural design and draughting has replaced traditional practices completely. Computer literacy and proficiency in using architectural software applications are paramount (Doyle, 2019).

Due to the technical and vocational nature of the programme, physical infrastructural facilities are critical, and students are expected to practise during and after classes. Advanced technology that enhances transformations in the programme requires adequate space and equipment in order to enable quality teaching and learning. A multiplicity of studies undertaken to investigate the nexus between the curriculum, physical and information technology infrastructure and the degree of work-readiness of architectural technologists in Nigeria has been observed. For example, Sule *et al.* (2019) sought to elicit the perception of the role and job prospects of architectural technologists using a purposively selected sample of participants (undergraduates) from Kaduna Polytechnic.

In two related studies focusing on staff and students of departments of architecture in universities in Southeast Nigeria, accredited by the National Universities Commission (NUC), Ezeji (2016, 2017) recommended that CAD-literacy should be a key criterion for staff recruitment. Hamma-Adama and Kouidder (2018) assessed technological infrastructure, quality of human resources, and CAD opportunities available to graduates of built environment schools. Based on the study, it was concluded that architectural technology schools were leading in CAD training across Nigerian tertiary institutions with over 50% acquisition of the co-ordinated software, even though scarcities of human resources and technological infrastructure, in terms of shortages of skilled labour to train in BIM-related competencies, persisted. Similarly, Marcel-Okafor and Okafor (2021) investigated the contributions of the extant architectural technology towards improved BIM proficiency among architectural technologists as a pathway for improved awareness of sustainability. It was observed that 96% of the respondents had acquired such competencies, with 56% indicating that they had done so as students. Furthermore, Marcel-Okafor (2017) explored the impact of the architectural technology curriculum on the practical performance of graduates working in the industry. Gumau and Osunkunle (2010) articulated the challenges facing architectural technologists in contemporary Nigerian society. It was found that these challenges culminate from an extensive training period at the polytechnics and the inadequacy of accredited institutions to train prospective technologists, particularly in Northeast Nigeria, among others (Gumau & Osunkunle, 2010)

Whilst it is evident that studies about the impact of the curriculum and physical and information technology infrastructure on the degree of work-readiness of architectural technologists in the Nigerian context abound, mono-methods for data collection and analysis were used in most of these studies (Marcel-Okafor & Okafor, 2021; Hamma-Adama & Kouidder, 2018; Ezeji, 2016, 2017; Gumau & Osunkunle, 2010). Thus, the dearth of mixed methods design studies to support data has hindered the presentation and application of inclusive insights which might be imperative for resolving any challenges. Hence, the choice of this design provided an opportunity to achieve this goal in this study.

Research Methodology

An explanatory sequential mixed methods research design was adopted. This design involved the collection of quantitative data in the initial phase and the collection of qualitative data in the second phase. Data collected from the two phases were analysed independently, thereby enabling the use of the results from the latter to explain and interpret the findings of the former (Creswell & Plano Clark, 2007).

Accordingly, a survey questionnaire was used to obtain data from the respondents comprising academic staff, architectural technologists in practice, and the architectural firms spread across the five states that constitute Southeast Nigeria, namely, Abia, Anambra, Ebonyi, Enugu, and Imo. The second phase comprised semi-structured interviews carried out within a multi-case study setting. The selection criteria for the cases and participants are outlined in the subsequent sections.

Primary data were obtained from the questionnaires and interviews in a sequential manner, related to the following:

1 Perceptions of architectural firms of the competence of graduates in accomplishing the expected learning outcomes of the HND programme, and the proficiency of graduates in using computer software applications.
2 Availability and adequacy of infrastructural facilities used in training the technologists in architectural technology departments of the polytechnics.
3 Views of technologists in practice pertaining to the relevance of professional courses learned in school to the demands of practice.

Sample and Sampling Techniques

In the initial phase, samples for this study were derived from three research populations, namely, five polytechnics within Southeast Nigeria that offer NBTE-accredited architectural technology programmes, HND architectural technology graduates of the institutions who practised within the study area, and employers from architectural firms that employed these technologists within the study area.

Owing to the heterogeneous structure of ownership of the polytechnics, stratified random sampling was adopted in the ratio of 1:2 to define the samples of this research population (Kenton, 2019). Hence, two homogenous groups comprising two state-owned institutions and three federal-owned institutions were identified. However, one of the two state-owned polytechnics opened a comparatively young department of architectural technology and, thus, was not considered in the selection. Hence, one state-owned and two federal-owned polytechnics were selected randomly for the study by balloting. A total of 37 copies of a structured questionnaire, based on a pilot survey, were administered to the academic staff in the three departments that constituted the entire sample size for the polytechnics.

The results of a pilot survey revealed that a total number of 358 HND students graduated from the departments over a period of five years. This formed the population size (N) for the second research population. However, not all the graduates worked in firms within the study area. Hence, an assumption of 50% of the total number of HND graduates (179 graduates) was used to determine the population size (N), by applying Yamane's simplified formula as follows:

$$n = \frac{N}{1 + N(e)^2} \qquad (10.1)$$

where,

'n' is the sample size,
'N' is the population size, and
'e' is the level of precision.

$$= \frac{179}{1+179(.05)^2} = \frac{179}{1.4475} = 123.661 \approx 124, \text{ therefore } n = 124 \text{ (Sample size).}$$

An allowance of 5% was made for envisaged low response. This culminated in a total of 130 questionnaires being administered to the HND graduates in practice across the five states that constituted the study area. The instrument was administered to the graduates from the three selected polytechnics on a first-contact basis according to a list of graduate members registered with the respective state chapters of the Nigerian Institute of Architects (NIA).

The results of a pilot survey also showed an uneven spread of 75 registered architectural firms located within the study area, with Enugu State and Abia State having the highest and lowest number of firms registered with Architects Registration Council of Nigeria (ARCON) respectively. However, Ebonyi State had no firm registered with ARCON at the time that documentation was obtained from the register of architectural firms compiled by the Council (ARCON, 2016). To determine the sample size (n) for this third research population, a sample size for the study was adopted based on desired accuracy with a population percentage or variability of 50%, a confidence level of 95%, and a 5% margin of error (Taherdoost, 2016, 2017). Hence, 63 ARCON registered architectural firms were adopted as the sample size for the study. However, owing to the uneven spread of the firms across the study area, a proportional allocation was used to determine the number of firms to be assessed within each case study, and a simple random sampling method, i.e. quota sampling, was used to select the number of samples from the derived population (Gill *et al.*, 2010).

Interview Samples

The academic staff and graduates practising in firms within the study area participated in the follow-up explanations during semi-structured interviews conducted within a 12-week window in the second phase. The interviews were conducted in the schools and firms located in the states listed in Table 10.1. The primary researcher and two pre-trained research assistants conducted the interviews on each occasion which lasted for an average of 30 minutes. The interviews were recorded with the participant's informed consent, transcribed by the research assistants, and coded using QDA Miner Lite 4.0 Software.

Data Collection

Validity, reliability, and practicability of the instruments were checked using a pilot survey and summarised in Table 10.2.

Table 10.1 Total number of interviews per state

Polytechnics	*Abia State Polytechnic*	*Federal Polytechnic Oko, Anambra State*	*Federal Polytechnic Nekede, Imo State*
Academic staff in departments	11	14	15
States in which firms are located	Abia State	Anambra State	Imo State
Principal architects in architectural firms	4	6	14

Source: Original.

Table 10.2 Specific principles and strategies adopted in the initial phase

Rigour principles	*Strategies*	*Application to study*
Validity	Content-related evidence: Survey method	Random sampling process was adopted in the selection of the sample size. Experienced lecturers scrutinised research instruments.
Reliability	Internal consistency method	Questionnaire tested using pilot survey. Respondents were selected on first contact.
Practicability	Administration of Instruments	Direct method of administration.

Source: Original.

A direct method of administering the structured questionnaire was used in the initial phase. The research assistants administered the questionnaire to the respondents on a first-contact basis and maintained professional cordiality with the respondents.

In the second phase, quantitative results were obtained from the follow-up interviews involving semi-structured pre-arranged questioning guided by approved themes and conducted consistently and systematically. Eight open-ended questions were designed to gain further understanding of the staff and graduates concerning the phenomenon being studied.

Data Analysis

The quantitative data collected were analysed at multiple levels using the Statistical Package for Social Sciences (SPSS) Version 26. The data collected were analysed at three levels: the uni-variate level which involved frequency distribution analysis using descriptive techniques. The second level was a bi-variate analysis, in which the correlations amongst the variables were established. The third and final level was a multi-variate analysis, in which regression analysis was carried out. This process of analysis is summarised in Table 10.3. The data collected was presented using tables, graphs, and charts to enhance visual appreciation of the information garnered.

Table 10.3 Summary of instruments for QUAN data analysis

Phase primacy	*Research questions*	*Related H_0 and variables to test*	*Measurement scale*	*Analysis to test H_0*	*Other analyses*
Phase 1 QUAN	**RQ1:** To what extent have learning outcomes/objectives of the programme influenced the abilities of the graduates working in architectural firms in Southeast Nigeria?	$H_0 1$ *Practical performance* versus *proficiency in CADD*	Ordinal variables	Spearman Rho correlation analysis	Frequency distribution analysis
	RQ2: To what extent have the physical infrastructural facilities in the schools affected the abilities of the graduates working in architectural firms in Southeast Nigeria?	$H_0 2$ *Available functional computer workstations* versus *practical performance*	Interval variables	Pearson Product Moment correlation	Frequency distribution analysis

Source: Original.

Table 10.4 Summary of instruments for QUAL data analysis

Phase primacy	*Research questions*	*QUAL data collection*	*QUAL data analysis*
Phase 2 QUAL	**RQ3**: Are the professional courses and entrepreneurial skills taught relevant to the dynamics of practice in Southeast Nigeria?	Multiple case studies from three randomly selected polytechnics and states. Semi-structured interviews.	Qualitative content analysis. Induced inferences.

Source: Original.

The qualitative data collected were fully transcribed and analysed using inductive coding in content analysis, in which identified excerpts of the data are fitted to a set of pre-determined codes using QDA Miner Lite 4.0 Software, as summarised in Table 10.4.

Presentation and Analysis of Findings

In this part of the study, the analysis of the data collected is reported, using techniques indicated in the research methodology, and the corresponding findings that ensued. It is emphasised that these data were derived from the research questions which addressed the topic, aim, and objectives of this study.

Summary of Quantitative Data Analysis

On a five-point Likert Scale, ranging from "poor" to "excellent", the respondents were requested to rate the graduates' performances relating to the stipulated learning outcomes of HND programme objectives, as outlined in the NBTE Education Course Curriculum and Specification.

Analysis according to the study area of aggregated data on *"graduates' ability to assist in design and prepare area diagrams, circulation and overall spatial relationship"* showed that most of the graduates investigated were given good performance ratings in this ability, as illustrated in Figure 10.1.

Analysis of aggregated data on *"graduates' ability to develop and prepare details of design drawings"* showed predominantly positive ratings for graduates employed, as illustrated in Figure 10.2.

Analysis according to the study area of *"graduates' ability to read and interpret technical drawings and specifications and use details in working drawings"* revealed that most of the graduates employed were given good performance ratings in this ability, as shown in Figure 10.3.

Analysis according to the study area of the data on *"graduates' ability to prepare spatial programmes and working drawings for projects"* showed that one-third of the graduates of federal polytechnics, Nekede and Oko, who were employed, were given either fair or poor performance ratings in their ability to perform this task. The majority had positive ratings, as illustrated in Figure 10.4.

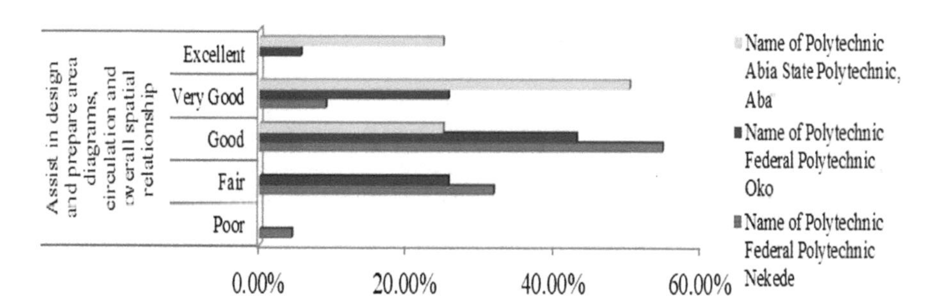

Figure 10.1 Area-wise data on graduates' ability to assist in designing and preparing area diagrams, circulation, and overall spatial relationship

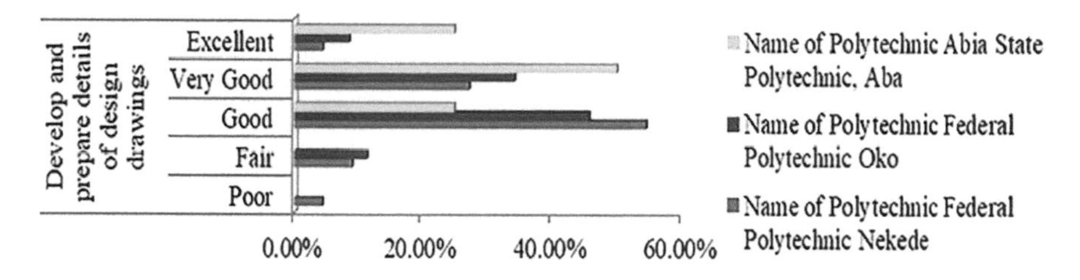

Figure 10.2 Area-wise data on graduates' ability to develop and prepare details of design drawings

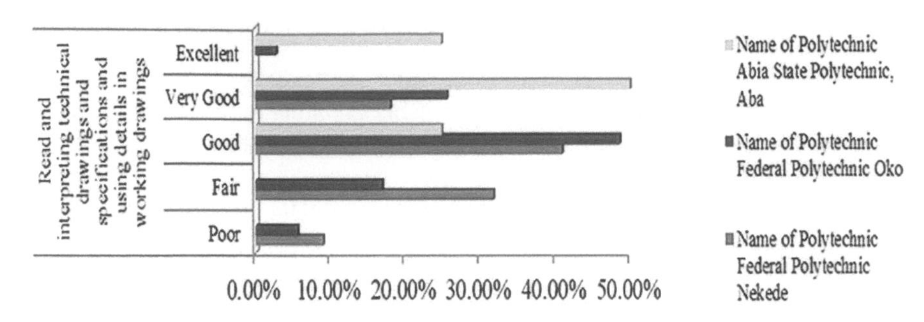

Figure 10.3 Area-wise data on graduates' ability to read and interpret technical drawings and specifications and use details in working drawings

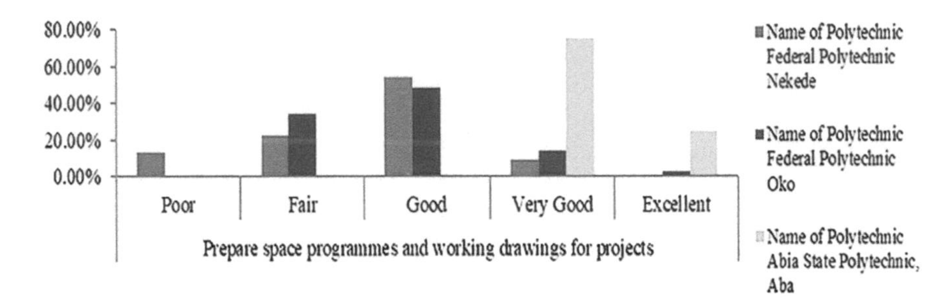

Figure 10.4 Area-wise data on graduates' ability to prepare spatial programmes and working drawings for projects

Analysis of aggregated data on *"graduates' ability to detail building components in wood, plastic, metal and concrete"* showed that over half of the graduates of federal polytechnics, Nekede and Oko, as well as the entire proportion of graduates of Abia polytechnic, who were employed, were given either good or very good performance ratings in their ability to accomplish this task, as illustrated in Figure 10.5.

Analysis of aggregated data on *"graduates' ability to make presentation drawings and detailed perspectives of architectural works, with necessary rendering"* revealed that most employed graduates were given positive performance ratings of good, very good and excellent, as illustrated in Figure 10.6.

Analysis according to the study area of "*graduates' proficiency in using AutoCAD"* revealed that on a three-point rating scale ranging from "none" to "expert proficiency", most HND graduates had basic proficiency in the use of this software app and almost one-third of the graduates displayed expert proficiency, as illustrated in Table 10.5.

Analysis according to the area of *"graduates' overall performance in using ArchiCAD"* showed that over one-third of the graduates displayed expert proficiency in using this software, as illustrated in Table 10.6.

Analysis according to the area of *"graduates' overall performance in using Revit"* showed that close to half of the graduates displayed basic proficiency in using Revit, as illustrated in Table 10.7.

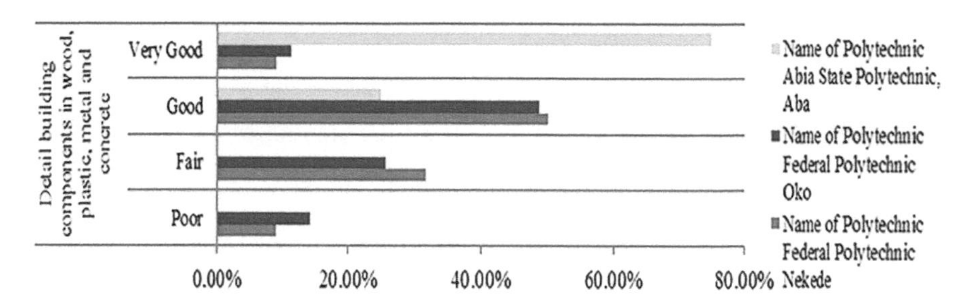

Figure 10.5 Area-wise data on graduates' ability to detail building components in wood, plastic, metal, and concrete

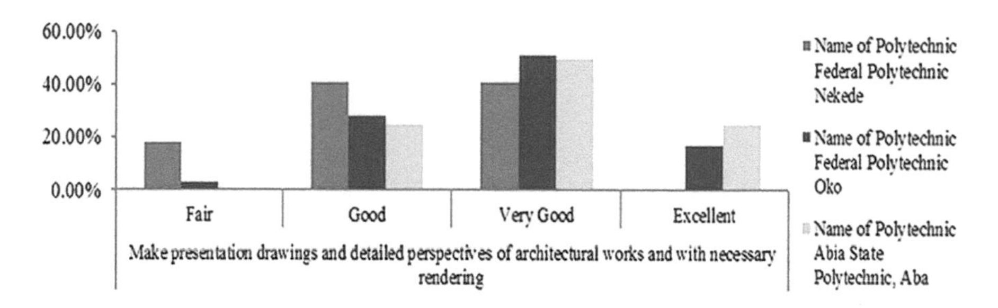

Figure 10.6 Area-wise data on graduates' ability to make presentation drawings and detailed perspectives of architectural works, with necessary rendering

Table 10.5 Area-wise data on graduates' proficiency in using AutoCAD

% within Name of Polytechnic Federal Polytechnic Nekede		*Name of Polytechnic*		
		Federal Polytechnic Oko	*Abia State Polytechnic, Aba*	
Proficiency in using AutoCAD	None	9.1%	5.7%	
	Basic proficiency	68.2%	62.9%	25.0%
	Expert proficiency	22.7%	31.4%	75.0%
Total		100.0%	100.0%	100.0%

Source: Original.

Table 10.6 Area-wise data on graduates' overall performance in using ArchiCAD

% within Name of Polytechnic Federal Poly. Nekede		*Name of Polytechnic*		
		Federal Poly. Oko	*Abia State Poly. Aba*	
Proficiency in using ArchiCAD	None	13.6%		
	Basic proficiency	50.0%	60.0%	25.0%
	Expert proficiency	36.4%	40.0%	75.0%
Total		100.0%	100.0%	100.0%

Source: Original.

Table 10.7 Area-wise data on graduates' overall performance in using Revit

% within Name of Polytechnic Federal Poly. Nekede		*Name of Polytechnic*		
		Federal Poly. Oko	*Abia State Poly, Aba*	
Proficiency in using Revit	None	31.8%	11.4%	
	Basic proficiency	36.4%	51.4%	75.0%
	Expert proficiency	31.8%	37.1%	25.0%
Total		100.0%	100.0%	100.0%

Source: Original.

Table 10.8 Regression coefficients table for *practical performance of graduates*

	Coefficients				
	Standardised coefficients				
	Beta	*Bootstrap (1,000) estimate of std. error*	*df*	*F*	*Sig.*
Proficiency in using AutoCAD	0.178	0.117	2	2.318	0.110
Proficiency in using ArchiCAD	0.209	0.107	2	3.809	0.029
Proficiency in using Revit	0.112	0.147	2	0.585	0.561
Relevance of entrepreneurial skills to present work demands	0.354	0.103	1	11.890	0.001

Source: Original.

The results of categorical regression coefficients (Table 10.8) revealed that the independent variables, *proficiency in using ArchiCAD* and *relevance of entrepreneurial skills to present work demands* were found to have significant correlation with *practical performance of graduates*.

Summary of Qualitative data analysis

Theme 1: Competencies Enabling Employability
Responses to: "*Style of studio draughting in the firm in which you practice?*"
All the graduates said that they worked with various software packages, and the firms provided only computer workstations. Most of the interviewees used ArchiCAD, closely followed by AutoCAD users. Some used other applications not mentioned in the study such as Adobe Photoshop, SketchUp, and Artlantis.

Responses to: "*Criteria for employment in firm?*"
The graduates said that their certificate and CAD proficiency were paramount criteria for being employed in the firm.

Theme 2: Satisfaction with Course Curriculum and Delivery
Responses to: "*Relevance of course contents in Architectural Design Courses?*"

Most of the graduates believed that the contents of the design courses had been beneficial to their practice, although they were also of the opinion that the courses required updating.

Responses to: *"Relevance of course contents in computer courses?"*
The graduates said that the contents of computer courses provided the basic knowledge. A majority said their proficiency increased outside school.

Theme 3: Availability of infrastructure
Responses to: *"Adequacy of software packages used for computer courses taught in the HND Programme?"*
Most of the interviewees said that the software packages were not adequate and advocated that more software packages should be used in teaching computer courses.

Responses to: 'Availability of drawing tables and stools?'
Majority of the interviewees said that these were not adequate.

Responses to: *"Availability of modelling studios and equipment for entrepreneurial skills?"*
Most of the graduates said that there were adequate modelling studios and computer rooms fully equipped, although many computers were not functional.

Responses to: *"Availability of electricity supply and internet facilities in the department?"*
Most of the interviewees said that these were available, although alternative supply from generators was used frequently.

Integration of Quantitative and Qualitative Findings

On a design level, the choice of follow-up explanatory method provided comprehensive insights into the effects of physical infrastructural facilities on the practical performance of graduates within the study area. This was necessary because of the paucity of mixed methods research design studies on the subject and the multi-dimensional factors that affected the practical performance of graduates (Oyedele & Tham, 2007).

Quantitative and qualitative findings were integrated to explicate, illustrate, and validate data further to generate robust conclusions. Data integration was achieved in the study example by adopting a follow-up explanations model, which enables integration at the levels of design, methods and interpretation (Fetters *et al.*, 2013). In the follow-up explanatory model, integration was done at the interpretation level, as the results of respective analyses of the two datasets were integrated. Table 10.9 shows the joint display of QUAN and QUAL results (Follow-up model

Implications of the Study

This research design provided the framework to create inclusive combined datasets. The quantitative data collected and analysed provided a broad perspective of the research problem, and the results of the qualitative data analysed provided further explanation (Creswell & Plano Clark, 2018). The integration at the interpretation level provided the connection of findings from the initial QUAN phase with QUAL findings in the second phase (Berman, 2017). The implications of the results revealed a need to introduce more computer software packages in the departments and subsequent regular training of the staff in appreciation and application of the packages. The presentation of findings in a joint display (Table 10.9) using a follow-up model, a variant of Explanatory Design in which qualitative data were used to explicate quantitative results, helped to assimilate results visually from both datasets and identify new insights (Fetters *et al.*, 2013).

Conclusions

In this study, the application of an explanatory sequential mixed method design was illustrated in assessing the effects of physical infrastructural facilities in schools on the practical performance of architectural technology graduates of polytechnics in Southeast Nigeria. The quantitative results showed that proficiency in using ArchiCAD and relevance of entrepreneurial skills to present work demands were found to have a significant correlation with the practical performance of graduates. Qualitative findings provided further details on criteria for employment, satisfaction with curriculum and delivery and availability of infrastructure. The results of QUAL confirmed the findings of QUAN.

This research design provided a robust approach to the study to garner inclusive data and discover the strengths of the design. Findings revealed independent variables that had a significant correlation with the dependent variable in QUAN analyses. This was expounded further by findings in QUAL analyses. However, the limitations identified included initial reluctance to respond to survey instruments. This was checked by providing proof of confidentiality and academic purpose of their responses. Based on the study, the introduction of more computer software packages in the departments and subsequent regular training of the staff in appreciation and application of the packages was recommended.

References

ARCON. 2016. *Register of Architects Entitled to Practice in the Federal Republic of Nigeria.* Architects Registration Council of Nigeria, Abuja.

Armstrong, G. & Allwinkle, S. 2017. Architectural technology: The technology of architecture. In: M. Aurel (ed). *51st International Conference of the Architectural Science Association*, pp. 803–812. The Architectural Science Association and Victoria University of Wellington, New Zealand.

Armstrong, G., Comiskey, D. & Pepe, A. 2013. Architectural technology: A brave new design? *4th International Congress of Architectural Technology (ICAT).* Sheffield Hallam University, Sheffield, UK.

Berman, E.A. 2017. An exploratory sequential mixed methods approach to understanding researcher's data management practices at UVM: Integrated findings to develop research data services. *Journal of eScience Librarianship*, 6(1): 1–24.

Creswell, J.W. & Plano Clark, V.L. 2007. *Designing and Conducting Mixed Methods Research.* Sage Publishing, Cincinnati.

Table 10.9 Joint display of QUAN and QUAL results (Follow-up model)

QUAN data collection	*QUAN data analysis*	*QUAN results*	*Identify results for follow-up*	*QUAL data collection*	*QUAL data analysis*	*QUAL Findings*	*Interpretation QUAN – QUAL*
Survey method using structured questionnaire	**For RQ1:** Spearman Rho correlation analysis and **For RQ2:** Pearson Product Moment correlation	*Proficiency in using ArchiCAD* and *relevance of entrepreneurial skills to present work demands* were found to have a significant correlation with *practical performance of graduates*	Relevance of course contents Availability of facilities	Semi-structured interviews	**Qual. content analysis**	The graduates are employed based on certificates and CADD proficiency Design courses learned in school were beneficial to practice. Computer courses learned provided basic proficiency	Confirmation/ Expansion

Source: Original.

Creswell, J.W. & Plano Clark, V.L. 2018. *Designing and Conducting Mixed Methods Research.* 3rd edition. Sage Publications, Thousand Oaks, CA.

Doyle, A. 2019, July 23. *Important Job Skills for Architects.* Retrieved August 2020, from the balancecareers: www.thebalancecareers.com

Emmitt, S. 2002. *Architectural Technology.* Blackwell Science Ltd., Oxford.

Emmitt, S. 2009. *Architectural Technology.* John Wiley & Sons, London.

Ezeji, K.E. 2016. Implementation of computer-aided design curriculum in National Universities Commission accredited architecture departments in South-East Nigeria. *Unpublished PhD thesis*. Chukwuemeka Odumegwu Ojukwu University, Anambra State, Nigeria.

Ezeji, K.E. 2017. Correlation of computer aided design proficiency and curriculum content amongst Architecture students in Universities in South-East Nigeria. *Journal of Teaching and Education*, 6(02): 281–292.

Fetters, M.D., Curry, L.A. & Creswell, J.W. 2013. *Achieving Integration in Mixed Methods Designs - Principles and Practices.* Retrieved from Pubmed.gov: doi:10.1111/1475-6773.12117. Epub.

Gill, J., Johnson, P. & Clark, M. 2010. *Research Methods for Managers.* 4th edition. SAGE Publications, London.

Gumau, W.S. & Osunkunle, A. 2010. Travails of the Polytechnic architectural technologist in Nigeria. *Journal of Research in Education and Society*, 1: 2.

Hamma-Adama, M. & Kouidder, T. 2018. A quest needs for Building Information Modelling Tools Training in a Developing Nation. *7th International Congress on Architectural Technology (ICAT 2018)*, pp. 87–105. Robert Gordon University, Aberdeen.

Kenton, W. 2019, August 22). *Investopedia.* Retrieved October 14, 2019, from www.investopedia.com: https://www.investopedia.com

Maina, J.J. & Salihu, M.M. 2016. An Assessment of generic skills and competencies of architecture graduates in Nigeria. *AJETS*, 9(1): 30–41.

Marcel-Okafor, U.O. 2017. Curriculum and methods of teaching architectural technology in Nigerian polytechnics: Challenges and implications on practice performance of graduates. *International Journal of Informatics*, 10(4): 1410–1417.

Marcel-Okafor, U. & Okafor, M. 2021. Enhancing building information modelling (BIM) training in Nigerian polytechnics: Towards sustainable development in Southeast Nigeria. *E3S Web of Journals*, 295: 05020.

National Board for Technical Education. 2016. *Brief History of National Board for Technical Education (NBTE).* Retrieved April 27, 2017, from www.nbte.gov.ng

Oladapo, A.A. 2017. The impact of ICT on Professional practice in the Nigerian construction industry. *The Electric Journal of Information Systems in Developing Countries*, 24: 17.

Olotuah, A.O., Taiwo, A.A. & Ijatuyi, O.O. 2016. Pedagogy in architectural design studio and sustainable architecture in Nigeria. *Journal of Educational and Social Research*, 6(2): 157–164.

Opoko, A.P. & Oluwatayo, A.A. 2015. Architectural Education for today's challenges. *Arts and Design Studies*, 38: 24–30.

Oyedele, L.O. & Tham, K.W. 2007, June 30. *Building and Environment.* Retrieved May 30, 2017, from www.elsevier.com/locate/buildenv.

Sule, B.E, Muhammad, S.U., Diso, A.U., Toluhi, J. & Musa, A. 2019. Perception of the role and job prospects of architectural technology graduates of Kaduna polytechnic. In: Oluigbo, S.N. & Sagada, M.L. (eds). *Proceedings of the 2019 Association of Architectural Educators in Nigeria (AARCHES)*. School of Post-Graduate Studies, Ahmadu Bello University, Zaria.

Taherdoost, H. 2016. Sampling methods in research methodology; How to choose a sampling technique for research. *International Journal of Advance Research in Management*, 5(2): 18–27.

11 An Exploration of Sustainable Procurement Practice in Irish Construction-Contracting Firms

Duga Ewuga, Mark Mulville, and Alan Hore

Summary

It has been argued that sustainable procurement can promote collaboration and innovation. However, despite the calls made to construction firms to adopt and implement sustainable procurement practices, most construction firms seem to be slow to embrace such practices. An exploratory sequential mixed methods approach was used to explore the organisational strategies adopted by large construction-contracting firms in Ireland in driving the adoption and implementation of sustainable procurement. This approach provided the advantage of triangulating the data to corroborate the research findings of the study. The intention of this chapter was to illustrate and guide researchers in using the mixed methods research design to understand organisational strategies. Also, the study is an example of how a case study research strategy engenders the adoption of an exploratory sequential mixed methods research design. Despite some limitations, the mixed methods approach enabled an in-depth exploration of complicated research questions concerning how large construction-contracting firms address sustainability issues in their procurement process. It also enabled the researchers to collect a richer and more robust array of evidence that cannot be accomplished by any single method alone.

Introduction

To drive the practice of sustainability, the government of the Republic of Ireland has developed various strategies and incentives. Some of these initiatives include the declaration of the climate emergency (Government of Ireland, 2019), the net-zero energy building (NZEB) (SEAI, 2017) and an increase in public funding for energy-efficient buildings from €35 million in 2018 to over €100 million (Irish Green Building Council [IGBC], 2018). However, despite these efforts by the government, reports about the sustainable progress index (SPI) have shown that Ireland is ranked 15th among 15 European Union countries regarding environmental issues (Social Justice Ireland, 2021). The construction industry has been identified as one of the main contributors towards the prevalence of unsustainable production and consumption patterns in the country. The procurement process remains one of the facets that can be used to curtail this trend. This observation has made an

DOI: 10.1201/9781003204046-11

investigation into the implementation of sustainable procurement practices within construction-contracting firms imperative.

Exploring the sustainable procurement practices of construction-contracting firms in the Republic of Ireland requires a back-and-forth approach. The back-and-forth enables constant switching between empirical observations and theory, which generates a greater understanding of both empirical phenomena and theory (Dubois & Gadde, 2002; Upstill-Goddard *et al.*, 2016). The advantage of switching between methods is being able to collect data from multiple sources such as documents, surveys and interviews to understand the research problem better (Dubois & Gadde, 2002; Yin, 2009; Upstill-Goddard *et al.*, 2016).

The focus of this chapter is on a demonstration of how the exploratory, sequential, mixed methods research design was used in exploring sustainable procurement practices of the top 50 construction-contracting firms in the Republic of Ireland.

Sustainable Procurement in Construction-Contracting Organisations

The Architectural, Engineering and Construction (AEC) sector has been the most significant global consumer of raw materials and accounts for about 25%–40% of the world's total carbon emissions (World Economic Forum, 2016). In an effort towards driving sustainability, global business leaders and civil society have made declarations to lead their organisations towards achieving the sustainability goals and objectives (Business and Sustainable Development Commission, 2017). At the sectoral level, it has been noted that construction firms have made commitments to implement sustainability practices (Berry & McCarthy, 2011; Meehan & Bryde, 2011; Zuo *et al.*, 2012). In their organisational policies and mission statements, sustainability has been positioned as a primary focus. However, it is noted that these policies are developed by most construction firms only to seek legitimacy and competitive advantage rather than truly intending to implement the practice of sustainability (Rietbergen *et al.*, 2015; Upstill-Goddard *et al.*, 2015; Russell *et al.*, 2018).

Based on earlier studies, Akotia *et al.* (2016) argued that there is a disparity between the theoretical concept and reality regarding sustainability in the AEC sector. Furthermore, Upstill-Goddard *et al.* (2016) revealed that most small and medium-sized firms in the AEC sector always adopt a reactive stance to the adoption of sustainability. Such challenges could likely be because of firms not identifying or understanding the benefits to be derived from implementing such practice (Mulligan *et al.*, 2014; Ruparathna & Hewage, 2015; Upstill-Goddard *et al.*, 2016). Other reasons that have been deduced as contributing to the low level of sustainability adoption in the AEC sector include the complex structure of the industry and the uniqueness of construction projects that involve various actors and a supply chain.

The procurement mechanisms of organisations have been identified as a driver for the adoption and implementation of sustainability practices (Bratt *et al.*, 2013; Grob & Benn, 2014). Sustainable procurement in construction is the process or mechanism whereby organisations or firms collaborate with their various supply chains and relevant stakeholders to achieve the sustainability goals of a project (Berry & McCarthy, 2011). Therefore, there is a need to explore further how construction firms collaborate with their supply chains and other organisational resources to achieve the sustainable procurement goals of their organisations. Exploring the sustainable procurement practices of large construction-contracting firms is vital because they are often regarded as being a learning model for other firms (Chang *et al.*, 2016).

Research Methodology

In this section, the research strategy adopted for the study is explained, including the choice of research methodology, the case study and unit of analysis, the sampling method, and data collection and analysis techniques.

The Case Study Strategy

To achieve the aim of the research, it is necessary to select a suitable strategy. The choice of a research strategy depends on the type of research question posed and the extent of control that an investigator has over the actual behavioural events (Yin, 2009). A case study research strategy was found to be appropriate to explore how construction-contracting firms utilise their organisational resources to drive sustainable procurement. The case study strategy involves investigating a relatively small number of naturally-occurring (rather than researcher-created) cases (Hammersley, 1992). The benefit of using a case study for construction research lies in its reliability of capturing rich information for the investigation by providing the researcher(s) with the opportunity to retain the holistic and meaningful characteristics of real-life events (Barrett and Sutrisna, 2009).

Using a mixed methods research design in a case study, Yin (2009) suggested two, nested arrangements. The first approach is a case study within a survey. The primary research relies on a survey or other quantitative techniques for data collection to study the main case. While the other approach involves a survey within a case study. In this approach, the primary research relies on holistic data but surveys or other quantitative techniques are used to collect data about the case.

To understand the sustainable procurement practices of the top 50 construction-contracting firms, the second approach, being a survey within a case study, was found to be appropriate. The survey within a case study was appropriate because the aim of the study was to provide an inference of the best explanation rather than to generalise the findings. Studying a single construction-contracting firm can provide a good insight into representing a typical case of other large construction-contracting firms. Therefore, data from the top 50 construction-contracting firms was collected using a quantitative questionnaire survey. In contrast, the qualitative data was obtained from documents and interviews involving a single construction-contracting firm.

Case and Unit of Analysis

The case and unit of analysis have often been considered to be the same (Baxter & Jack, 2008; Yin, 2009). However, Fernie (2005) explained that the definition of the case differs, although not necessarily, from the unit of analysis. Fernie (2005) explained that the case is a broad outline of the objectives of the study, while the unit of analysis is used to define the level at which the object will be studied. The research question and the proposition are connected to the unit of analysis because they are used to guide and identify the relevant information to be collected about the case (Yin, 2009). Therefore, for this study, the unit of analysis was defined as the sustainable procurement practice associated with the top 50 construction-contracting firms in the Republic of Ireland, whilst the case was based on the selected, single construction-contracting firm.

Document analysis and semi-structured interviews were used to obtain data to understand the sustainable procurement practice of a single selected firm, whilst the questionnaire survey was used to collect data from respondents who were undertaking various roles in the top 50 construction-contracting firms, in a sequential manner.

The Mixed Methods Research Design

The choice of exploratory, sequential mixed methods design was deemed to be appropriate for exploring the sustainable procurement practice of the top 50 construction-contracting firms. However, it is noted that mainstreaming sustainability in the AEC sector is not a uniform linear process (Boyd & Schweber, 2012). Therefore, exploring the sustainable procurement practices of construction-contracting firms in the Republic of Ireland requires a back-and-forth approach.

Sampling Technique

Deciding on the sample size in mixed methods research is fairly complicated because the sampling technique must be designed for the qualitative and quantitative research components of the study (Onwuegbuzie & Collins, 2007; Teddlie & Yu, 2007; Creswell, 2014). The number of participants in a study could be determined by various means, whether through *a priori* determination or a more adaptive approach such as saturation (Sim *et al.*, 2018). Onwuegbuzie and Collins (2007) explained further that determining sample size for a mixed methods study should be informed by the research goal, objective, purpose and question(s). The choice of the sampling technique (i.e. random or non-random) is based on the type of generalisation of interest, which could be a statistical generalisation or analytical generalisation (Onwuegbuzie & Collins, 2007; Denscombe, 2014). Another critical factor in choosing the sampling technique for a mixed methods study is the time orientation (sequential or concurrent) and the purpose of mixing the quantitative and qualitative data. Such purposes could include triangulation, complementarity, initiation, development or expansion (Onwuegbuzie & Collins, 2007; Teddlie & Yu, 2007; Creswell, 2014).

Finally, the last step to be considered is the relationship between the qualitative and quantitative samples. Onwuegbuzie and Collins (2007) explained that these relationships could be identical, parallel, nested or multi-level. An identical relationship indicates that an equal number of sample members is drawn in the quantitative and qualitative phases, whereas a parallel relationship specifies that the sample size for qualitative and quantitative components of the research is different but is drawn from the same population of interest. The nested relationship implies that the sample members selected for one phase of the study represent a subset of those chosen for the other phase of the investigation. Lastly, the multi-level relationship involves using two or more samples extracted from different levels of the study (i.e. different populations).

Therefore, in defining the sample size for this study, the researchers reflected on the research question and the purpose of the study. Figure 11.1 presents the considerations influencing the choice of sampling technique adopted for the study.

Overall, purposive sampling (non-random) was selected as the sampling technique of choice for the study based on the considerations highlighted in Figure 11.1.

Data Collection

In undertaking this research, multiple techniques and procedures for data collection were adopted. As explained in the Research Methodology Section, qualitative data was collected from a single construction-contracting firm in the first instance and, subsequently, quantitative data was elicited from representatives of the top 50 construction-contracting firms using questionnaires. In exploring the various strategies for

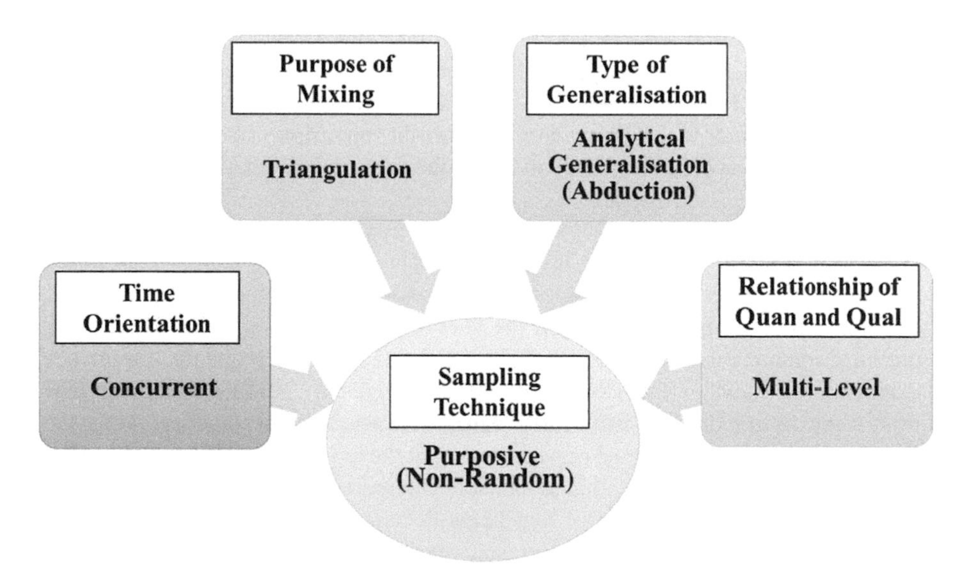

Figure 11.1 Considerations influencing the choice of sampling technique for the study

Source: Original (2022)

Table 11.1 Role of respondents

Role	*Frequency*	*Percentage (%)*
Managing Director	7	11
Regional Director	7	11
Director	7	11
Commercial Manager	15	24
Contracts Manager	15	24
Sustainability Manager	1	2
Strategy and Business Development Manager	1	2
Procurement Manager	5	8
Chief Estimator	4	6
Total	62	100

Source: Original.

driving sustainable procurement, the organisational policies and management procedures of the single construction-contracting firm were reviewed. Additionally, further data was obtained from a semi-structured interview with a representative of the single construction-contracting firm. The level of importance and performance attached to the various strategies for driving sustainable procurement was examined during the semi-structured interview. Finally, to corroborate the findings of the qualitative data, a questionnaire with a five-point Likert Scale was designed to evaluate the level of importance and performance relating to the various strategies for driving sustainable procurement at the top 50 construction-contracting firms.

Data was collected from purposively selected senior members of the organisations, who were involved in the management of projects and procurement. A total of 62 responses were received from the questionnaire survey, as shown in Table 11.1. Four interviews were conducted with commercial managers (2), a contracts manager (1) and a senior estimator (1) of the single construction-contracting firm selected. In compliance with ethical considerations, the transcripts of the interviews were sent to the interviewees for affirmation before the commencement of the analysis.

Data Analysis

The documents and interview transcripts were analysed using thematic analysis (Braun & Clarke, 2006), whereas data from the questionnaire survey were analysed using principal component analysis (PCA) (Tabachnick & Fidell, 2013) and the Importance Performance Analysis (IPA) (Rial *et al.*, 2008; Chang *et al.*, 2017), respectively. The PCA was used to reduce the factors to a manageable size for better analysis, whilst the IPA provided a view of the level of performance of the top 50 construction-contracting firms in driving and implementing sustainable procurement practices.

The IPA is a managerial tool that helps decision-makers to set management priorities and determine how scarce resources might be allocated best (Rial *et al.*, 2008; Chang *et al.*, 2017). IPA is based on the mean performance and mean importance obtained from a surveyed respondent for several attributes or characteristics of a specific factor, product or service. Performance refers to the level of satisfaction with an attribute, while importance refers to the attributes being

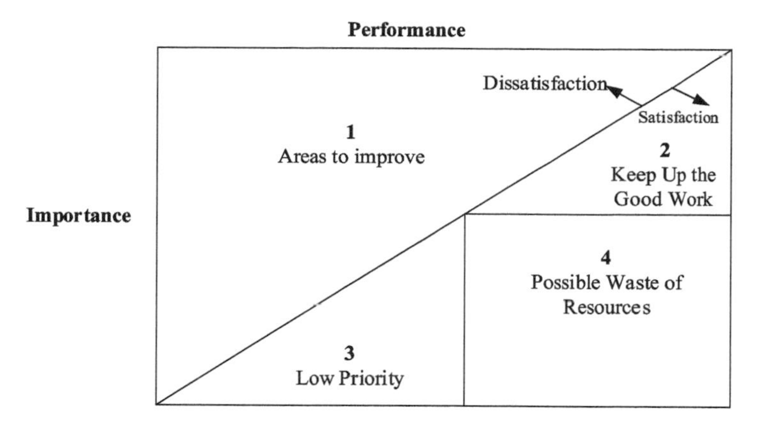

Figure 11.2 Revised IPA

Source: Rial et al. (2008)

assessed by the respondents (Martilla & James, 1977). The average values of importance and performance of different attributes were calculated concerning one another, mainly in the area divided into four quadrants. The horizontal axis represented performance, while the vertical axis represented the importance, as shown in Figure 11.2.

Depending on which quadrant a particular attribute is located, managers can decide which attributes are the top priorities and low priorities for improvement. As shown in Figure 11.2, the quadrants are

i **Areas to improve (Q1):** attributes in this quadrant are perceived to be highly important while the performance levels are relatively low. Such attributes explain that immediate corrective action or efforts should be concentrated on improving performance.

ii **Keep up the good work (Q2):** attributes in Q2 are perceived to be of high importance, and the company or enterprise has high-performance levels on these attributes. This represents the strengths and competitive advantage of companies whose task is to continue maintaining the quality of the elements contained in the quadrant.

iii **Low priority (Q3):** attributes in this quadrant are of low importance, and the enterprise has low-performance levels. Such attributes represent no threat to the organisation.

iv **Possible waste of resources (Q4):** attributes in this quadrant are of low importance levels, but the enterprise has high-performance levels on these attributes. This indicates that the organisation is spending valuable resources on minor elements.

Presentation of Results

The findings from the study are presented in this section.

Findings from Analysis of Documents

Findings from the analysis of the selected firm's sustainability and sustainable procurement policies revealed that sustainability was an essential issue in the company's procurement process. As part of the sustainable procurement policy, it is stated that

> We recognise and accept that our actions can have consequences on others and we will strive to ensure that decisions taken regarding the procurement and engagement of any organisations, individuals, goods or services are governed by integrating environmental, legal, social and economic considerations into all stages of the procurement process.

Regarding how the environmental, social and economic considerations will be integrated, it is stated in the sustainability policy that

> …the policy will be communicated across the organisation and will be regularly reviewed and updated….

Furthermore, it is stated in the procurement policy that

> We will seek to collaborate with clients, consultants, specialist contractors and suppliers to develop integrated supply chains which respect biodiversity and human rights and promote fair employment practice.

Findings from Interviews

How the firm used various organisational strategies in collaborating with their clients, designers and supply chains to promote sustainable procurement was explored during the interviews. The interviewees explained that their organisational policies and management procedures guided them in addressing their clients' needs and government regulations. For instance, collaborating to meet the government laws and regulations, a commercial manager explained that

> In this country, we have the Environmental Protection Agency (EPA). So, they manage to make sure that, if we are starting a site, there is a watercourse beside us. We get in touch with them to ensure we are not polluting the watercourse. (CM1).

Likewise, the interviewees explained that they usually educate their clients on alternative construction methodologies and sustainable products and materials. Another commercial manager (CM2) explained how they educate their clients and revealed that, as a company, they have different initiatives that guide them in suggesting alternative products. For example, the commercial manager explained that

> If a client specified a one-ply roofing system, we will say, no, we are not doing this; we need to a have two- or three-ply system. This is because, if the project fails, it is ours for nearly forever until the defects liability period is finished. (CM2).

Similarly, an Estimator (ES1) revealed further that, depending on the type of client, when they are bidding for a job, they usually provide alternative products/materials. Providing such alternatives provides the client(s) with several options. The estimator stated that

> It depends on the client and, if we feel that the client favours sustainability, we go for more sustainable products. We would put that in our technical submission, so the client knows it. So, if they see that some items or prices are more expensive, they know that we are sourcing from a better location, or our products are LEED certified. (ES1).

Regarding collaboration amongst staff in the organisation, the interviewees explained that the company management procedure had made it compulsory for all staff to participate in specific training. Some of the training they are expected to participate in regarding sustainability includes bribery, child labour, anti-slavery, and health and safety. Also, during the delivery of a project, the interviewees explained that the project is broken into different packages that enable each team to have a full understanding of a particular trade or package. For example, a contracts manager (ConM1) stated that

> Each team member became a specialist like an engineer. He could just be focused only on the façade so that he will have enough time to review and share his knowledge with the opposite partners around or with other colleagues on the opposite side like the design team. (ConM1).

Lastly, regarding collaborating with supply chains, the interview findings revealed that the main contracting firm had developed various strategies to ensure that their supply chains meet the sustainability requirements of the company. Also, they had developed a high level of trust and long-term relationships. One of the interviewees (CM1) stated that

> It might be an advantage to us to have a subcontractor who would kind of have gone with you on the journey for the better job for the clients. Also, the subcontractors we would use and re-use would be somebody who would make us look good. (CM1).

Another strategy noted was that the main contracting firm provides an opportunity for the lower-tier supply chains to move to a higher tier. The firm provides this opportunity by supporting the lower supply chain with various resources and training. An Estimator (ES1) explained that

> What we do is that we would give a sub-contractor a job, and we would say, okay, we are going to give you this contract. It is bigger than what you are used to getting, but we will help you with it. We will give you a lot more support in getting the job done and administratively as well. Our site team would lend a hand. At least then we can bring that sub-contractor to the next level. (ES1).

Finally, the interviewees revealed that the company organises a supply chain award apart from providing support to their supply chains. Such an award is given to the supply chain organisation that performed well in the various company projects. The interviewees believed that such an award provides the supply chains with the opportunity to improve their social capital.

The findings from the documents and interviews showed that the firm studied was mindful of its image and actions. Therefore, the firm had developed various strategies for collaborating with their clients, supply chains, staff and other stakeholders. In the next phase of the study, using the questionnaire survey, the performance of the various strategies of the top 50 construction-contracting firms in the Republic of Ireland was examined.

Results from Questionnaire Survey

Powmya *et al.* (2017) identified 22 strategies for driving sustainable procurement to be considered by construction organisations. As shown in Table 11.2, these strategies are grouped into human resources, technology, finance, knowledge, capacity development and environmental pro-activeness.

PCA was used to explore the performance of the top 50 construction-contracting firms in implementing the 22 strategies. The PCA results grouped the 22 factors into three groups, as shown in Table 11.3 and Figure 11.3. The groups were

Table 11.2 Implementation strategies

Function	*Strategies*
• Human resources	• Recruitment of experienced technical staff • Education and training • Employee empowerment, and • Employees reward system
• Technology	• Improvement of communication system through information technology
• Finance	• Surety • Bonds, and • Insurance policies
• Knowledge	• Monitoring and Evaluation of projects • Inter-firm collaboration, and • Continual professional development
• Capacity development	• Collaboration with international sustainable construction body • Collaboration with international bodies • Collaboration with international sustainable construction firms • Collaboration with various size of contractors • Partnering with suppliers, and • Research and development
• Environmental pro-activeness	• Compliance with sustainability legislation • Voluntary rating and environmental management system (EMS) • Industrialised building systems (IBS)/fabricated building units • Sustainable procurement, and • Sustainable construction management

Source: Powmya *et al.* (2017).

classified into three (core, enabling and supplemental capabilities) (Barton, 1995). The core capabilities are the knowledge and skills that reside in the organisation, such as technical expertise and staff skills. These capabilities are the source of competitive advantage while enabling capabilities are necessary but not sufficient to distinguish a company competitively. In other words, enabling capabilities are those which a firm must have in support of its normal operations and core capabilities. Finally, supplemental capabilities add value to core capabilities, but they could be imitated.

In the next phase of the analysis, the level of importance-performance of the firms regarding the 22 strategies identified was examined. The mean scores of performance and importance were determined. The gap (discrepancies) of each of the strategies was also determined. The gap was determined by subtracting the mean scores of performance factors from the mean scores of important factors. Tables 11.4–11.6 show the IPA results for the three groups of capabilities identified.

Examining the core capabilities in Table 11.4, the results indicated that most firms were performing well in all of the strategies identified except for C1 (improving communication systems through information technology). These results

Table 11.3 Total variance explained

Component	*Initial eigenvalues*			*Extraction sums of squared loadings*			*Rotation sums of squared loadings*
	Total	*% of variance*	*Cumulative %*	*Total*	*% of variance*	*Cumulative %*	*Total*
I	11.868	53.947	53.947	11.868	53.947	53.95	9.140
2	1.739	7.906	61.853	1.739	7.906	61.85	7.454
3	1.432	6.510	68.363	1.432	6.510	68.36	7.868
4	0.927	4.216	72.579				
5	0.824	3.744	76.322				
6	0.691	3.139	79.462				
7	0.565	2.568	82.030				
8	0.526	2.393	84.423				
9	0.516	2.346	86.769				
10	0.463	2.107	88.875				
11	0.378	1.718	90.593				
12	0.350	1.591	92.184				
13	0.280	1.273	93.457				
14	0.260	1.182	94.639				
15	0.250	1.138	95.776				
16	0.194	0.880	96.657				
17	0.178	0.811	97.468				
18	0.175	0.797	98.264				
19	0.126	0.573	98.838				
20	0.104	0.471	99.309				
21	0.099	0.449	99.758				
22	0.053	0.242	100.000				

Source: Original.

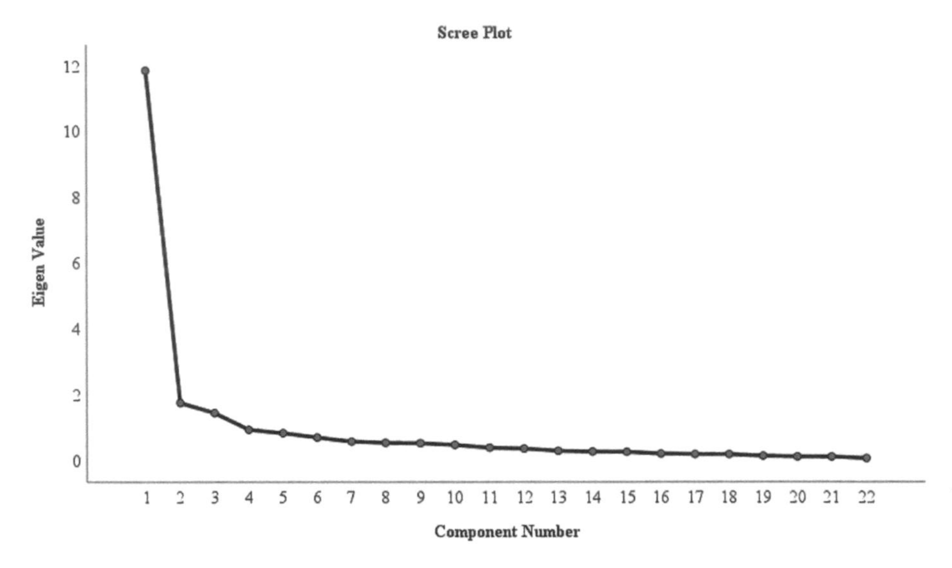

Figure 11.3 Scree plot

Source: Original

Table 11.4 Importance–performance analysis of core capabilities

Code	*Factor l: core capabilities*	*Mean of performance*	*Mean of importance*	*Gap*	*Focus*
Cl	Improving communication system through information technology	3.94	4.26	−0.32	Area to improve
C2	Employee reward system	3.45	3.44	0.02	Low priority
C3	Employee empowerment	3.74	3.89	−0.15	Low priority
C4	Education and training	4.19	4.39	−0.19	Keep up the good work
cs	Continual professional development	3.98	4.15	−0.16	Keep up the good work
C6	Inter-firm collaboration	3.48	3.63	−0.15	Low priority
C7	Recruitment of experienced technical staff	4 03	4.32	−0.29	Keep up the good work
	Grand total	3.79	3.88	−0.10	

Source: Original.

indicated that top construction-contracting firms would invest significant resources in developing their core capabilities.

With regard to enabling capabilities, as shown in Table 11.5, apart from item E2 (collaboration with international bodies), which required improvement, all other items were of low priority to most firms. A likely reason for the scores might be that enabling capabilities are not sources of competitive advantage, so firms will not prioritise addressing them.

Finally, Table 11.6 shows that the firms were performing well in developing the various strategies as supplemental capabilities. These capabilities support core

Table 11.5 Importance–performance analysis of enabling capabilities

Code	*Factor 2: enabling capabilities*	*Mean of performance*	*Mean of importance*	*GAP*	*Focus*
El	Collaboration with international sustainable construction firms	3.10	3.10	0.00	Low priority
E2	Collaboration with international bodies	3.19	3.34	−0.15	Areas to improve
E3	Collaboration with international sustainable construction body	3.34	3.48	−0.15	Low priority
E4	Collaboration with varying size contractors	3.52	3.40	0.11	Low priority
E5	Research and development	3.40	3.55	−0.15	Low priority

Source: Original.

Table 11.6 Importance–performance analysis of supplemental capabilities

Code	*Factor 1: supplemental capabilities*	*Mean of performance*	*Mean of importance*	*Gap*	*Focus*
Sl	Project and client requirement	4.37	4.52	−0.15	Keep up the good work
S2	Stakeholders' engagement	4.18	4.23	−0.05	Keep up the good work
S3	Compliance with sustainability legislation	4.18	4.34	−0.16	Keep up the good work
S4	Monitoring and evaluation of projects	4.16	4.26	−0.10	Keep up the good work
S5	Post-project evaluation and review	3.73	3.81	−0.08	Low priority
S6	Compliance with the voluntary rating and environmental management system (EMS)	3.89	4.00	−0.11	Keep up the good work
S7	Surety, bonds and insurance policies	3.97	3.87	0. 10	Possible waste of resources
S8	Partnering with suppliers	3.85	3.89	−0.03	Keep up the good work
S9	Industrialised building systems (IBS) prefabricated building units	3.71	3.76	−0.05	Low priority
S10	Collaboration amongst the various teams in your organisation	3.90	3.87	0.03	Possible waste of resources

Source: Original.

capabilities but they are not sources of competitive advantage because they can be imitated. Firms should pay attention to addressing such capabilities but, from the results in Table 11.6, Items S7 (surety, bonds and insurance policies) and S10 (Collaboration amongst the various teams in your organisation) were perceived as being a possible waste of resources. Therefore, it would be advantageous if management allocated resources to areas that need improvement.

Discussion of the Results

An exploratory, sequential, mixed methods research design was used to explore the sustainable procurement practices of construction-contracting firms in the Republic of Ireland. The convergent, parallel mixed methods design made it possible to collect and analyse the data almost simultaneously. In earlier studies, it was found that the adoption of sustainability in the AEC sector was low (Upstill-Goddard *et al.*, 2015; Akotia *et al.*, 2016; Upstill-Goddard *et al.*, 2016), and firms applied their procurement mechanism to drive their sustainability practice. Data were collected from documents, interviews and questionnaires to provide insight into an organisation's sustainable procurement practices. One of the benefits of using the mixed methods approach lay in being able to combine data to corroborate research findings (Saunders *et al.*, 2015).

Findings from the documents of the single, case study firm unveiled the company's sustainability policies, plans and intentions. Also, the interviews revealed how the policies were implemented and how collaboration was carried out amongst various team members and their supply chains. Furthermore, the interview findings revealed the various forms of training and support provided to staff and supply chains regarding the adoption and implementation of sustainability. The results from the questionnaire survey concurred with much of the findings from the documents and interviews. For example, the importance of integrating the policies across the organisation to their staff, supply chains and clients is stated clearly in the firm's sustainability policy. The level of importance and performance of such strategies, as observed from the questionnaire survey using Importance–Performance Analysis (IPA), showed that most of the firms were performing very well in most of the strategies, as seen in Table 11.4 (C4, C5 and C7) and Table 11.6 (S1–S3 and S8). These results indicated that large firms always develop strategies that will make them attain legitimacy and competitive advantage.

The study's findings provided a view of the performance of construction-contracting firms, enabling them to identify areas that require improvement in their practice of sustainability.

Implication for Practice and Research

From the findings of the study, it was evident that top construction-contracting firms have developed several strategies to drive their practice of sustainable procurement. These findings will guide firms that are struggling to adopt and implement sustainable procurement to develop their organisational strategies based on their available resources. Also, the findings provided a view of the areas that firms would have to improve and, areas where they were performing well. These will help them to allocate resources to achieve their goals and objectives for the practice of organisational sustainability. These results were obtained because of the use of exploratory, sequential, mixed methods design which made it possible to collect and analyse data almost simultaneously, enabling constant switching between empirical observations and theory. Doing so generates a greater level of understanding of both observed phenomena and theory (Dubois & Gadde, 2002; Upstill-Goddard *et al.*, 2016)

Conclusion

The purpose of this study was to demonstrate how an exploratory, sequential, mixed methods research design was used to explore the sustainable procurement practice of the top 50 construction-contracting firms in the Republic of Ireland. Using this research design and a case study strategy made it possible to explore the internal organisational strategies of construction-contracting firms. The approach enabled a deep investigation by examining a firm's sustainability policies (documents) and practice (interviews), as well as the views of other large firms through a questionnaire survey. The findings provided a view of the performance of the top 50 construction-contracting firms in the Republic of Ireland. Also, the findings revealed that most of the firms invested resources in developing their core and supplemental capabilities. Further study will be required to investigate areas that need improvement based on the findings of the Importance–Performance Analysis.

References

Akotia, J., Opoku, A., Egbu, C. & Fortune, C. 2016. Exploring the knowledge 'base' of practitioners in the delivery of sustainable regeneration projects. *Construction Economics and Building*, 16(2): 14–26.

Barney, J. 1991. Firm resources and sustained competitive advantage. *Journal of Management*, 17(1): 99–120.

Barrett, P. & Sutrisna, M. 2009. Methodological strategies to gain insights into informality and emergence in construction project case studies. *Construction Management and Economics*, 27(10): 935–948.

Barton, D.L. 1995. *Wellsprings of Knowledge: Building and Sustaining the Sources of Innovation*. Harvard Business School, Boston, MA.

Baxter, P. & Jack, S. 2008. Qualitative case study methodology: Study design and implementation for novice researchers. *The Qualitative Report*, 13(4): 544–559.

Berry, C. & McCarthy, S. 2011. *Guide to Sustainable Procurement in Construction*. CIRIA, London.

Boyd, P. & Schweber, L. 2012. Variations in the mainstreaming of sustainability: A case study approach. *Association of Researchers in Construction Management, ARCOM - Proceedings of the 28th Annual Conference,* Edinburgh, UK, 3–5 September 2012. https://www.arcom.ac.uk/-docs/proceedings/AR2012_Proceedings_Vol2.pdf

Bratt, C., Hallstedt, S., Robèrt, K.H., Broman, G. & Oldmark, J. 2013. Assessment of criteria development for public procurement from a strategic sustainability perspective. *Journal of Cleaner Production*, 52: 309–316.

Braun, V. & Clarke, V. 2006. Using thematic analysis in psychology. *Qualitative Research in Psychology*, 3(2): 77–101.

Business and Sustainable Development Commission. 2017. *Better Business Better World* [online]. Available at: http://report.businesscommission.org/uploads/BetterBiz-BetterWorld_170215_012417.pdf

Chang, R.-D., Zuo, J., Soebarto, V., Zhao, Z.-Y., Zillante, G. & Gan, X.-L. 2016. Sustainability transition of the Chinese construction industry: Practices and behaviors of the leading construction firms. *Journal of Management in Engineering*, 32(4): 05016009.

Chang, R.-D., Zuo, J., Soebarto, V., Zhao, Z.-Y., Zillante, G. & Gan, X.-L. 2017. Discovering the transition pathways toward sustainability for construction enterprises: Importance-performance analysis. *Journal of Construction Engineering and Management*, 4: 17.

Creswell, J.W. 2014. *Research Design: Qualitative, Quantitative, and Mixed Methods Approaches*. SAGE, Thousand Oaks, CA.

Crotty, M. 1998. *The Foundations of Social Research: Meaning and Perspective in the Research Process*. SAGE, London.

Denscombe, M. 2014. *The Good Research Guide: For Small-Scale Social Research Projects*. Mcgraw-Hill Education, Maidenhead.

Dubois, A. & Gadde, L.-E. 2002. Systematic combining: An abductive approach to case research. *Journal of business research*, 55(7): 553–560.

Easterby-Smith, M., Thorpe, R. & Jackson, P.R. 2015. *Management and Business Research*. SAGE, London.

Fernie, S. 2005. *Making Sense of Supply Chain Management in UK Construction Organisations: Theory Versus Practice*. Loughborough University, Loughborough.

Flick, U. 2007. *Designing Qualitative Research*. Sage Publications Ltd., Thousand Oaks, CA.

Government of Ireland. 2019. *Climate Action Plan 2019*. [Online]. Available at: https://assets.gov.ie/10206/d042e174c1654c6ca14f39242fb07d22.pdf [Accessed: 27 June 2021].

Grob, S. & Benn, S. 2014. Conceptualising the adoption of sustainable procurement: An institutional theory perspective. *Australasian Journal of Environmental Management*, 21(1): 11–21.

Guba, E.G. & Lincoln, Y.S. 1994. Competing paradigms in qualitative research. *Handbook of Qualitative Research*, 2(105): 163–194.

Hammersley, M. 1992. *What's Wrong with Ethnography? Methodological Explorations*. Psychology Press., London.

Irish Green Building Council (IGBC). 2018. *Creating an Energy Efficient Mortgage for Europe Building Assessment Briefing: IRELAND*. Energy Efficient Mortgages Action Plan initiative, Dublin.

Ketokivi, M. & Choi, T. 2014. Renaissance of case research as a scientific method. *Journal of Operations Management*, 32(5): 232–240.

Laryea, S. & Hughes, W. 2011. *Negotiating Access into Firms: Obstacles and Strategies of Conference*. Available at: centaur.reading.ac.uk

Martilla, J.A. & James, J.C. 1977. Importance-performance analysis. *Journal of Marketing*, 41(1): 77–79.

Meehan, J. & Bryde, D. 2011. Sustainable procurement practice. *Business Strategy and the Environment*, 20(2): 94–106.

Mitchell, A. 2018. A review of mixed methods, pragmatism and abduction techniques. *Electronic Journal of Business Research Methods*, 16(3): 103–116.

Mulligan, T.D., Mollaoglu-Korkmaz, S., Cotner, R. & Goldsberry, A.D. 2014. Public policy and impacts on adoption of sustainable built environments: Learning from the construction industry playmakers. *Journal of Green Building*, 9(2): 182–202.

Onwuegbuzie, A.J. & Collins, K.M.T. 2007. A typology of mixed methods sampling designs in social science research. *Qualitative Report*, 12(2): 281–316.

Powmya, A., Abidin, N.Z. & Azizi, N.S.M. 2017. Contractor firm strategies in delivering green project: A reviewof conference. *AIP Conference Proceedings, 2017*. Available at: aip.scitation.org

Rial, A., Rial, J., Varela, J. & Real, E. 2008. An application of importance-performance analysis (IPA) to the management of sport centres. *Managing Leisure*, 13(3–4): 179–188.

Rietbergen, M.G., van Rheede, A. & Blok, K. 2015. The target-setting process in the CO2 Performance Ladder: Does it lead to ambitious goals for carbon dioxide emission reduction? *Journal of Cleaner Production*, 103: 549–561.

Ruivo, P., Oliveira, T. & Neto, M. 2015. Using resource-based view theory to assess the value of ERP commercial-packages in SMEs. *Computers in Industry*, 73: 105–116.

Ruparathna, R. & Hewage, K. 2015. Sustainable procurement in the Canadian construction industry: Challenges and benefits. *Canadian Journal of Civil Engineering*, 42(6): 417–426.

Russell, E., Lee, J. & Clift, R. 2018. Can the SDGs provide a basis for supply chain decisions in the construction sector? *Sustainability (Switzerland)*, 10(3): 1–19.

Saunders, M., Lewis, P. & Thornhill, A. 2009. *Research Methods for Business Students*. 5th edition. Perntice Hall.

Saunders, M., Lewis, P. & Thornhill, A. 2015. Research methods for business students. In: Mark N.K. Saunders, Philip Lewis & Adrian Thornhill (eds). *Understanding Research Philosophy and Approaches to Theory Development*, pp. 122–161. Pearson Education, Harlow.

SEAI. 2017. *Nearly Zero Energy Building Standard.* [Online]. Available at: https://www.seai.ie/business-and-public-sector/standards/nearly-zero-energy-building-standard/ [Accessed: 14 May 2021].

Sim, J., Saunders, B., Waterfield, J. & Kingstone, T. 2018. Can sample size in qualitative research be determined a priori? *International Journal of Social Research Methodology*, 21(5): 619–634.

Social Justice Ireland. 2021. *Measuring Ireland's Progress - Sustainable Progress Index 2021*. [Online]. Available at: https://www.socialjustice.ie/content/publications/measuring-irelands-progress-sustainable-progress-index-2021 [Accessed: 06 March 2021].

Tabachnick, B.G. & Fidell, L.S. 2013. *Using Multivariate Statistics*. 6th edition. Pearson, Boston, MA.

Teddlie, C. & Yu, F. 2007. Mixed methods sampling: A typology with examples. *Journal of Mixed Methods Research*, 1(1): 77–100.

Upstill-Goddard, J.D., Glass, J., Dainty, A.R.J. & Nicholson, I. 2015. Analysis of responsible sourcing performance in BES 6001 certificates. *Engineering Sustainability*, 168(ES2): 71–81.

Upstill-Goddard, J., Glass, J., Dainty, A. & Nicholson, I. 2016. Implementing sustainability in small and medium-sized construction firms. *Engineering, Construction and Architectural Management*, 23(4): 407–427.

World Economic Forum. 2016. *Shaping the Future of Construction - A Breakthrough in Mindset and Technology*. [Online]. Available at: http://www3.weforum.org/docs/WEF_Shaping_the_Future_of_Construction_full_report__pdf

Yin, R.K. 2009. *Case Study Research: Design and Methods*. 4th edition. Sage, Thousand Oaks, CA.

Zuo, J., Zillante, G., Wilson, L., Davidson, K. & Pullen, S. 2012. Sustainability policy of construction contractors: A review. *Renewable and Sustainable Energy Reviews*, 16(6): 3910–3916.

12 Investigating the Accident Causal Influence of Construction Project Features Using Sequential Exploratory Mixed Methods Design

Patrick Manu

Summary

Construction project features (CPFs) include underlying accident causal factors that emanate from pre-construction decisions made by key project actors. They are organisational, operational and physical attributes of projects such as the nature of the project, method of construction, procurement method and level of construction and duration of the project. This chapter is focused on how a sequential exploratory mixed methods approach was used to address two research questions regarding the accident causal influence of CPFs: (1) how do CPFs influence accident occurrence; and (2) what is the extent of their impact on health and safety? The approach involved interviews with 11 construction professionals and a subsequent survey which yielded 187 responses. The purpose of this chapter was not to reiterate the published findings regarding these questions but, rather, to offer insights regarding the rational process that led to the selection of a mixed methods approach and, subsequently, the steps taken to operationalise the approach to address the research questions. In doing so, the chapter is positioned as a useful reference to guide other health and safety researchers in the selection and deployment of the sequential exploratory mixed methods research design to address similar or different research questions in construction accident causation.

Introduction

In spite of the socio-economic benefits derived from the construction industry, the industry has persistently been one of the worst industries in terms of occupational safety and health (OSH) in several countries (Health and Safety Executive (HSE), 2021; U.S. Bureau of Labor Statistics, 2021). Accidents are commonplace on construction sites. Over the years, improvement efforts have included studies aimed at identifying the causal factors in construction accidents in order that appropriate mitigation measures can be implemented (Suraji *et al.*, 2001; Haslam *et al.*, 2005).

In construction accident causation studies, causal factors, two key hierarchies of causal factors in construction accidents have been reported, namely, distal/root/originating factors and proximal factors (Suraji *et al.*, 2001; Haslam *et al.*, 2005). Studies by Suraji *et al.* (2001) and Haslam *et al.* (2005) relating to construction accident causation are considered to be seminal. Suraji *et al.* (2001) developed

DOI: 10.1201/9781003204046-12

the constraint-response model of construction accident causation which indicates distal and proximal causal factors. The proximal factors are those that can lead directly to accident causation while the distal factors are those that can lead to the introduction of these proximal factors in the construction process. Similar to the constraint-response model, is the ConCa Model by Haslam *et al.* (2005), in which distal factors are identified as originating influences and proximal factors as shaping factors. The originating influences are the high-level determinants of the existence, nature and extent of immediate accident causes and they include permanent works design, project management and client requirements (Haslam *et al.*, 2005). The shaping factors include site constraints, schedule constraints and housekeeping. In these construction accident causation studies and other OSH studies (Lingard *et al.*, 2015), the need to pay attention to distal/originating causal factors has been emphasised to ensure improvement in construction OSH. The need to pay attention to distal/originating accident factors is reinforced by the fact that they emerge from the pre-construction stage of project procurement where project participants have the greatest ability to influence OSH through pre-construction decision-making (Lingard *et al.*, 2015).

CPFs fall into this category of underlying accident causal factors as they emanate from pre-construction decisions made by key project actors, such as clients, designers and project management team, and influence accident occurrence during the construction phase. They are organisational, operational and physical attributes of projects such as the nature of the project, method of construction, procurement method, level of construction and sub-contracting, duration of the project, site restriction and design complexity (Manu *et al.*, 2014).

Despite the established significance of accident causation of originating accident influences, prior to the studies by Manu *et al.* (2014, 2017a, 2017b) not much empirical research had been focused on these features. Generally, in OSH studies and reports only passing references had been made to the accident causation role of CPFs, without focused, in-depth, empirical investigations of this accident causal phenomenon. As a result, there was a lack of detailed insight into (1) how CPFs influence accident occurrence (i.e. the mechanism/process by which CPFs influence accident occurrence); (2) the degree of the inherent potential of CPFs to influence accident occurrence (i.e. their potential/ability to cause accident/harm) and (3) the degree of their associated OSH risk (i.e. likelihood of accident occurrence) (Manu *et al.*, 2014, 2017a, 2017b). Against this background, Manu *et al.* (2014, 2017a, 2017b) reported the outcomes of empirical studies on the accident causal influence of CPFs. The fundamental research questions addressed in the studies were

1 How do CPFs influence accident occurrence?
2 What is the degree of potential of CPFs to influence accident occurrence? and
3 What is the degree of H&S risk posed by or associated with CPFs?

Details of the findings addressing the research questions 1–3 are given in Manu *et al.* (2014, 2017a, 2017b). While different methodological approaches specific to addressing the relevant research questions were reported in the individual publications, collectively, the above questions were addressed in the three publications by the use of an over-arching, sequential, exploratory, mixed-methods approach. The purpose of this chapter is to consolidate and present how a coherent, over-arching sequential exploratory mixed methods approach was applied to address the research questions. This methodological perspective of the entire study on the accident causal influence of CPFs has not been reported and, therefore, represents the theme of this chapter. The purpose of the chapter was not to reiterate the findings published by Manu *et al.* (2014, 2017a, 2017b). In the next section of the chapter, the conceptualisation of the research is presented. This is followed by a discussion of the rational process that informed the selection of a sequential exploratory mixed methods approach as well as the steps taken to operationalise the approach. A reflection on the use of the sequential exploratory mixed methods is subsequently given prior to concluding remarks.

Conceptualising the Accident Causal Influence of CPFs

In order to address the above research questions, it was imperative first to understand how CPFs influence accident occurrence in order to be able to determine the extent of their impact on OSH (i.e. the degree of their potential to influence accident occurrence and the degree of the OSH risks posed by or associated with CPFs). To understand how CPFs influence accident occurrence, an important reference point was accident causation models/theories given that they are developed essentially to explain how accidents occur in reality. Consequently, several accident causation models were reviewed (e.g. energy transfer models, individual/human models/theories and systems models [Manu *et al.*, 2017a]) with a view to developing a conceptual model of how CPFs influence accident occurrence.

Drawing on the systems models of accident causation, in particular, the Constraint-Response and the ConCA models (Suraji *et al.*, 2001; Haslam *et al.*, 2005) because of their utility in helping to understand the accident causal mechanism of CPFs, a conceptual model depicting how CPFs influence accident occurrence was developed. A simplified illustration of the conceptual model is shown in Figure 12.1, while a detailed illustration is given in Manu *et al.* (2017a).

The model shows that CPFs emanating from pre-construction decisions made by project actors (e.g. client, designer, project manager/client advisor) subsequently

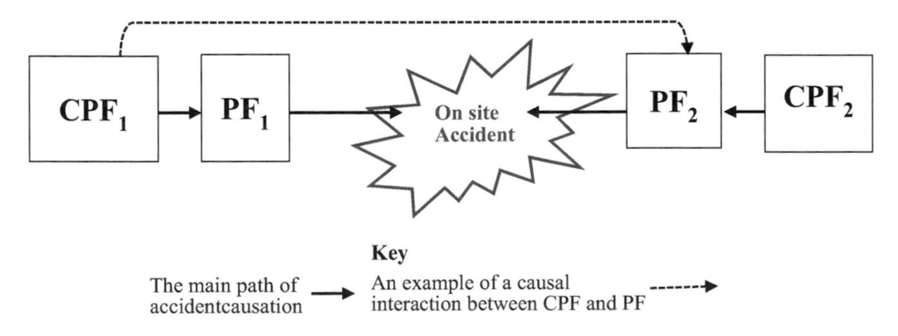

Figure 12.1 A simplified conceptual model of the accident causal influence of CPFs

induce proximate accident causes in the construction phase, which leads to accidents. For example, a complex design (a CPF arising from a client and/or designer decision) induces a proximate cause (i.e. difficulty in building – buildability) during onsite construction, which can eventually lead to an accident. Aside from this path of causation, there could be causal interactions between a CPF and the proximate causes induced by other CPFs. There could also be causal interactions between proximate causes. Such causal interactions could worsen or mitigate the extent of the impact of a CPF on OSH (Manu *et al.*, 2012).

Building on the above conceptualisation of how CPFs influence the occurrence of accidents (in response to Research Question 1, the next step was to determine an approach to assess the degree of the potential of CPFs to influence accident occurrence and the degree of OSH risk posed by or associated with CPFs, in order to address Research Questions 2 and 3, respectively. To achieve this, literature about evaluating OSH risk was reviewed. The review revealed that OSH risk evaluation falls into three main categories: qualitative; quantitative and semi-quantitative (Manu *et al.*, 2017b). Qualitative risk evaluation methods are easy to use, but subjective in nature and, therefore, can be difficult for a third party to understand the basis for the evaluation (WHO & FAO, 2009). Quantitative methods on the other hand, although they are more robust, are relatively difficult to apply and also require substantial historical and numerical data (BSI, 2008; Sachs & Tiong, 2009). A semi-quantitative approach provides a rigorous middle-ground between the quantitative and qualitative approaches, by mitigating the ambiguities associated with qualitative risk evaluation and the challenge of needing substantial data for quantitative risk evaluation. Semi-quantitative evaluation involves a risk matrix which combines scores for risk determinants (e.g. severity, exposure, likelihood). The scores are accompanied by qualitative descriptors, for example a score of "5" representing "very high likelihood" and a score of "4" representing "high likelihood". The research adopted semi-quantitative risk evaluation as a measurement framework to address Research Questions 1 and 2. To operationalise the semi-quantitative evaluation, a risk combination matrix based on the mathematical expression shown by Equation (12.1) was adopted. The equation was based on a widely used risk expression of *Risk = Hazard (i.e. potential of a thing to cause harm) x Exposure (i.e. exposure to the hazard)* (Duffus & Worth, 2001; Canadian Centre for Occupational Health and Safety, 2008). A detailed discussion of the derivation of the equation is given in Manu *et al.* (2017b).

$$\text{OSH risk associated with a CPF} = \text{The potential of the CPF to influence accident occurrence} \times \text{Exposure of workforce.} \quad (12.1)$$

Based on the above equation, for OSH risk associated with a CPF, it was necessary to determine its degree of potential to influence accident occurrence and exposure of the workforce to the CPF. For simplicity, exposure was taken to be a binary condition whereby, if a CPF was applicable to a project (e.g. if a project had a restricted site) then the workforce would be exposed to its potential to cause an accident but, if a CPF was not applicable to a project, then the workforce would not be exposed (Manu *et al.*, 2017b). Therefore, exposure was operationalised mathematically as:

"0" = workforce not exposed, where a CPF did not apply to a project; and "1" = workforce exposed, a CPF applied to a project. Consequently, the degree of OSH risk associated with a CPF on a project would be derived primarily from the degree of the potential of the CPF to influence accident occurrence. Therefore it was necessary for the empirical facet of the study to measure the degree of potential of a CPF to influence accident occurrence. The measure was then applied in a risk combination matrix based on Equation (12.1) to determine the degree of OSH risk associated with the CPF.

Research Methodology

Selection and Application of the Research Design

To aid the choice of an appropriate research design for the study, the tripartite framework for research design, developed by Creswell (2009), served as a useful guide. Creswell (2009) considered research design to be an intersection of three elements: paradigm, strategy of inquiry and research method. These elements are discussed next.

The Methodological Paradigm

Creswell and Creswell (2018: 5) used the term, "philosophical worldview" for "paradigm" and considered it to mean "a general philosophical orientation about the world and the nature of research that research brings to a study". Two prominent research paradigms are in common use: positivism and interpretivism (Fellows & Lui, 2015; Creswell & Creswell, 2018; Saunders *et al.*, 2019). According to positivism, it is assumed that a phenomenon obeys natural laws and can be subjected to quantitative logic while, according to interpretivism, it is assumed that a phenomenon does not obey natural laws but is interpreted based on peoples' conviction and/or understanding of the reality surrounding the phenomenon (Fellows & Lui, 2015; Creswell & Creswell, 2018; Saunders *et al.*, 2019). From the positivist perspective, reality can be observed independently as it is single and is therefore experienced in the same way by everyone. From the interpretivist perspective, reality can only be interpreted as it is multiple and is therefore experienced differently by everyone.

The research phenomenon under consideration and the key research questions influence the type of paradigm that has to be adopted (Pollack, 2007). Moreover, the conceptual model also forces the researcher to be rational and systematic about the constructs and variables to be included in the research instrument (Ahadzie, 2007). From the research questions posed in the introduction section, it is evident that they were laden with measurement and, therefore, in order for objective measurements to be obtained, it was logical to adopt positivism as an over-arching worldview for the phenomenon being investigated (i.e. the accident causal influence of CPFs). By adopting positivism, the degree of the potential of a CPF to influence accident occurrence and its associated degree of OSH risk could be viewed as a "single reality", which could then be observed and assessed objectively. The developed conceptual model, which was a prior formulation regarding the mechanism by which CPFs influence accident occurrence, also aligned well with the adoption of positivism.

Research Strategy and Method

In addition to the philosophical position adopted in research, researchers also adopt a research strategy and specific research methods for collecting and analysing data. The research strategy (i.e. strategy of inquiry) provides a specific direction for procedures in a research design (Creswell, 2009). The three common research strategies are qualitative, quantitative and mixed-methods strategies.

Qualitative research provides a means of exploring and understanding the meaning that individuals or groups ascribe to a phenomenon (Creswell & Creswell, 2009; Fellows & Liu, 2015). It is useful in answering research questions relating to "how" and "why" (Fellows & Liu, 2015). The qualitative process of research is inductive in relation to theory and literature and it is usually rooted in the interpretivist paradigm (Fellows & Liu, 2015; Creswell & Creswell, 2018). Examples of qualitative strategies include ethnography, grounded theory, case study, phenomenological research and narrative research.

Quantitative research is a means of testing objective theories or prior formulations by examining the relationship among variables. It involves numerical and objective measurements to address questions. It is thus useful in answering research questions relating to "what", "how much" and "how many" (Fellows & Liu, 2015). The quantitative process of research is deductive in relation to theory and literature and it is usually rooted in the positivist paradigm (Fellows and Liu, 2015; Creswell and Creswell, 2018). It involves the formulation of hypotheses or prior formulations in the form of a conceptual model based on theory and literature with subsequent collection and analysis of data to verify those prior formulations. Examples of quantitative strategies include surveys and experiments.

Mixed methods research is an amalgam of qualitative and quantitative strategies in a single study (Creswell & Creswell, 2018; Saunders *et al.*, 2019). Therefore it involves the use of both qualitative and quantitative methods of data collection and analysis in a single study (Creswell & Creswell, 2018; Saunders *et al.*, 2019). Mixed methods research is normally appropriate in research programmes where, because of the nature of the research problem being investigated, it is possible to collect both qualitative and quantitative data, the analysis of which would offer a better and deeper understanding of a phenomenon (Creswell & Creswell, 2018; Saunders *et al.*, 2019). Mixed methods strategies include sequential mixed methods and concurrent mixed methods. The sequential approach is two-fold: (1) sequential exploratory strategy, which begins with a qualitative strategy (e.g. phenomenological study) for exploratory purposes, followed by a quantitative strategy (e.g. survey) and (2) sequential explanatory strategy, which begins with a quantitative strategy (e.g. a survey) followed by qualitative strategy (e.g. phenomenological study). With the concurrent mixed methods approach, the investigator applies both qualitative and quantitative strategies concurrently during the study and then integrates or merges the information in the analysis and interpretation of the overall results.

Selecting and Applying the Sequential Exploratory Mixed Methods Strategy

Given that quantitative research is usually rooted in the positivist paradigm (Fellows & Lui, 2015; Creswell & Creswell, 2018; Saunders *et al.*, 2019) which was the paradigm adopted for the study, the quantitative strategy naturally emerged as a main strategy of inquiry for the research. The suitability of quantitative inquiry for answering questions relating to "what", "how much" and "how many" (i.e. measurement) further reinforced its suitability for this research given that the research questions largely suggested measurement. Again, the wish to have a generalised view regarding the degree of the potential of CPFs to influence accident occurrence and their associated OSH risk (which could inform pre-construction OSH decision-making) aligned well with the quantitative strategy as it is appropriate for making generalisations. In the main, the quantitative research strategy thus appeared to be a prime strategy for delivering the research.

Despite the suitability of the quantitative approach for the research, the need to verify how CPFs influence accident occurrence also introduced an element of qualitative inquiry given that such inquiry is more useful for answering questions relating to "how" and "why" (i.e. meaning) (Fellows & Liu, 2015). Although the conceptual model was a prior formulation, it essentially attempted to explain a phenomenon (i.e. how CPFs influence accident occurrence) and, as such, its verification could be achieved through a qualitative inquiry. The need mainly to measure the OSH impact of CPFs (implying quantitative inquiry), combined with the need also to seek an explanation regarding how CPFs influence accident occurrence (implying a qualitative inquiry) in a single study, led to an overall, mixed methods research strategy.

Since the conceptual model explaining how CPFs influence accident occurrence was instrumental in shaping the measurement framework (i.e. the procedure to measure the degree of potential of CPFs to influence accident occurrence and the degree of OSH risk associated with CPFs, as described above), it was logical that prior verification of the conceptual model (through a qualitative inquiry) preceded the deployment of the measurement framework (in a quantitative inquiry) so that any eventual refinement of the elements of the conceptual model could be incorporated in the deployment of the measurement framework. In view of this, the sequential exploratory mixed methods strategy emerged as the most appropriate mixed methods strategy for the research. Given that the initial qualitative phase of a sequential exploratory strategy enables a researcher to develop an instrument to be administered subsequently to a sample of the population (Creswell & Creswell, 2018), this strategy proved useful in developing the instrument used as part of operationalising the measurement framework. Building on the preceding discussion, the overall research design, therefore, entailed a sequential exploratory mixed methods approach underpinned by the positivist paradigm, which meant that the entirety of the subject of inquiry (i.e. how CPFs influence accident occurrence and their impact on OSH) was viewed as a single reality. While the pragmatist paradigm is usually associated with mixed methods approach (Creswell & Creswell, 2018), other paradigms could be adopted for mixed methods research depending on which aspect of the mixed approach is dominant, as well as the overall purpose of the research (Ankrah, 2007). Within construction management research, for instance, Ankrah (2007) adopted the sequential exploratory mixed method approach rooted in the objectivist/positivist paradigm in a study about culture in construction. Interviews were used initially in the study to verify a conceptual formulation and to help design an instrument (i.e. questionnaire), which was subsequently applied in a quantitative survey to assess the impact of cultural factors on project performance. Figure 12.2 shows an outline of the sequential exploratory mixed method approach as conceived and applied in the study.

In accordance with the sequential exploratory mixed methods approach, a qualitative inquiry preceded a quantitative inquiry. The qualitative inquiry involved a phenomenological study operationalised by the use of semi-structured interviews with professionals in construction. The phenomenological approach was used because it was deemed to be suitable for the exploration of professionals' opinions, knowledge and lived experiences of the causal influence of CPFs on accidents, with the intention of verifying/eliciting the key elements of the conceptual model (which were the main path of accident causation and the causal interaction). An interview schedule was developed to guide the interviewing process. Using a business directory, 50 contractors operating in the United Kingdom were selected randomly and invited by letter to participate in the interviews. In the invitation letter, contractors were requested to provide a professional in a construction management

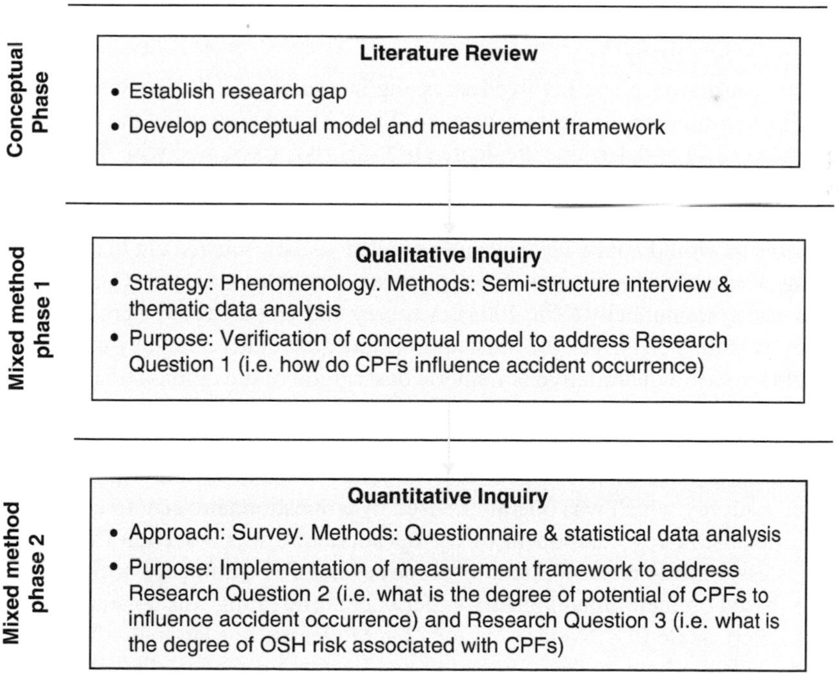

Figure 12.2 Overall research plan

role (e.g. H&S manager, project manager, construction manager or site manager) to participate in the interview. This was to ensure that the relevant professionals with the appropriate expertise were interviewed. In addition, industry contacts were used to enhance participation. Through the invitation letters and industry contacts, a total of 11 participants were recruited and interviewed face-to-face. An audio recorder was used to record the interviews. Out of the 11 interviews, saturation was reached after the ninth interview but, because prior interview arrangements had already been made with two other professionals, the additional two interviews were conducted. Analyses of the interviews entailed: transcription of the audio interviews; organisation and preparation of the transcripts; iterative re-reading of the transcripts; deductive and subsequent inductive coding of the interview transcripts; and theme generation. In the main, the findings of the qualitative inquiry further highlighted the undeniable existence of the causal influence of CPFs on accidents, as reported in the literature. Beyond the CPFs that were already reported in literature as influencing accident occurrence, the interviews also revealed "restriction of site locality" as another CPF with accident implications. The interview revealed that this project feature induces "difficulty in traffic (vehicle and pedestrian) control around site vicinity" as a proximate accident cause which could lead to accidents during the construction phase. Regarding how CPFs influence accident occurrence, the interview findings gave support and credence to the key features of the conceptualised view of how CPFs contribute to accident occurrence. The conceptual model was thus deemed to be a sound explanation of how CPFs influence accident occurrence. A detailed account of the interview and findings are provided in Manu *et al.* (2017a).

The quantitative phase involved assessing/measuring the degree of the potential of CPFs to influence accident occurrence (Facet 1) and then using the measures in Equation (12.1) to determine the degree of OSH risk associated with the CPFs. As previously mentioned, two prominent research strategies used in quantitative research are experiments and surveys. In choosing between these, it was resolved that experiments would not be ideal because they are usually carried out in a laboratory setting where the investigator can manipulate variables of interest directly, precisely and systematically (Yin, 2003). A survey and, in particular, a cross-sectional survey, was thus chosen as the most appropriate quantitative strategy of inquiry as it could provide a quantitative or numeric description of the opinion of a population based on a study of a sample of that population. The use of a survey thus meant it was possible to capture the opinions of construction professionals regarding the accident causal influence of CPFs. The first facet of the quantitative phase thus involved a survey which was operationalised by a questionnaire sent to construction professionals. A key component of the questionnaire was a section that required the professionals to rate the extent to which several CPFs influence accident occurrence (based on their broad industry experience). The rating was done using a five-point Likert Scale (0 = None, 1 = Low, 2 = Moderate, 3 = High, 4 = Very High). Again, with the help of the aforementioned business directory, the questionnaire was administered to 1,000 construction companies randomly selected from the business directory with a request for relevant construction professionals to complete the survey. The survey yielded 187 useable responses which were analysed to generate quantitative measures of the potential of the CPFs to influence accident occurrence. The survey responses were coded numerically in Microsoft Excel and subsequently exported into the Statistical Package for Social Sciences Software (SPSS) and R Software for statistical analyses including descriptive statistics (e.g. mean, mode and standard deviation) and inferential statistics (i.e. inter-rater agreement analysis). A detailed account of the survey and findings is reported in Manu *et al.* (2014). The quantitative measures were subsequently used as input data for Equation (12.1) to determine the OSH risk associated with the CPFs. The results indicated that CPFs, such as demolition, underground construction, tight project duration, high-level construction and multi-layer sub-contracting, are associated with a high OSH risk. Other CPFs such as traditional on-site construction, new work, design and build procurement, pre-assembly construction and single-layer sub-contracting are associated with a medium OSH risk. Details of the results are reported in Manu *et al.* (2017b).

Reflections

This section contains reflections on the merits/benefits and challenges regarding the use of the sequential exploratory mixed methods approach for the research with a view to sharing some lessons for other OSH and construction management researchers. A key benefit of the sequential exploratory mixed methods strategy is that it enables a researcher both to generalise findings to a population and to develop a detailed view of the meaning of a phenomenon. The research achieves these, firstly, by exploring generally in a qualitative manner to learn about what variables to study and, then, studying those variables with a large sample of individuals quantitatively (Creswell & Creswell, 2018). Consequently, the sequential exploratory mixed methods strategy can help a researcher to develop an instrument to be administered subsequently to a sample of the population (Creswell & Creswell, 2018). This feature of the sequential exploratory mixed methods strategy was particularly useful in this research in that, not only did the qualitative phase provide empirical verification of the conceptual model, but also it revealed an additional CPF (i.e. restriction of site locality) beyond those reported in the literature. The new CPF was incorporated into the design of the survey instrument. Without the initial qualitative inquiry (as part of the sequential, exploratory, mixed-methods strategy) the research would have missed this benefit.

However, the use of the sequential exploratory mixed methods is, not without challenges. A key challenge associated with the approach is its time-demanding nature (Creswell & Creswell, 2018). Since the qualitative and quantitative phases must be undertaken sequentially (and not concurrently), the duration of an entire study is prolonged and can be further aggravated if the qualitative phase is delayed or requires substantial time for data collection and analysis. This challenge was experienced during the study. The initial, qualitative phase took approximately five months for data collection and analysis (which entailed one pilot interview and 11 interviews) and the subsequent, quantitative phase took about seven months for data collection and analysis interviews (which entailed a pilot survey and a main survey). In order to mitigate the time-demanding nature of the sequential exploratory approach, two key actions were undertaken in this research: (1) concurrent interview data collection and analysis, such that interview data collection and analysis were undertaken in a start-to-start sequence; and (2) the design of the survey instrument commenced about mid-way during the qualitative phase rather than waiting until the completion of the analysis of all the interviews. These actions helped to redeem time and mitigate the adverse time impact of the sequential, mixed methods approach.

Conclusions

In construction management studies, including OSH research, a single methodological approach has commonly been applied (i.e. either a qualitative or quantitative approach). Recognising that both approaches have inherent limitations and that the limitations inherent in any single approach could help to mitigate the limitations of another, built environment researchers (including those in the construction OSH domain) ought to embrace increasingly the idea of mixing research methods in a single study. Seeking to contribute towards an increased application of mixed methods approaches in built environment research, an example of the application of sequential exploratory mixed methods has been provided in this chapter to address research gaps in the construction OSH domain. In the chapter, insights have been shared regarding the knowledge gaps that necessitated the study; the research conceptualisation based on literature/theories; a detailed account of the thought process that led to the selection of the sequential exploratory mixed methods approach; how the approach was executed; and reflections on the merits and challenges associated with the approach.

In conclusion, sequential exploratory mixed methods can offer a robust mechanism for addressing multiple research questions related to interrogating the meaning and measurement of a phenomenon in a single study, especially where a deeper understanding of the meaning of the phenomenon is a pre-requisite (or key) to defining the concepts and/or variables to be measured. A sequential exploratory mixed method can be powerful in either: (1) providing preliminary confirmation/verification of concepts/variables drawn from literature; or (2) potentially unearthing new relevant variables that are not evident in literature; or (3) offering both 1 and 2. However, researchers applying this approach ought to take into consideration its time-consuming nature and take steps to mitigate it. Mitigation actions could be: (1) concurrent collection and analysis of data during the initial qualitative phase in start-to-start sequence and (2) commencing early design of the instrument for the quantitative phase as soon as relevant insights start to emerge from the qualitative phase.

References

Ahadzie, D.K. 2007. A model for predicting the performance of project managers in mass house building projects in Ghana. PhD Thesis, School of Engineering and the Built Environment, University of Wolverhampton.

Ankrah, N.A. 2007. An investigation into the impact of culture on construction project performance. PhD Thesis, School of Engineering and the Built Environment, University of Wolverhampton.

Canadian Centre for Occupational Health and Safety. 2008. *Risk Versus Hazards*. Canadian Centre for Occupational Health and Safety, Ontario.

Creswell, J.W. 2009. *Research Design: Qualitative, Quantitative, and Mixed Method Approaches*. 3rd edition. Sage Publications, Thousand Oaks, CA.

Creswell, J.W. & Creswell, J.D. 2018. *Research Design: Qualitative, Quantitative, and Mixed Method Approaches*. 3rd edition. Sage Publications, Thousand Oaks, CA.

Duffus, J. & Worth, H. 2001. *The Science of Chemical Safety Essential Toxicology – An Educational Resource*. International Union of Pure and Applied Chemistry, Research Triangle Park, NC.

Fellows, R. & Liu, A. 2015. Research methods for construction. 4th edition. Wiley Blackwell Publishing, Chichester.

Haslam, R.A., Hide, S.A., Gibb, A.G.F., Gyi, D.E., Pavitt, T., Atkinson, S. & Duff, A.R. 2005. Contributing factors in construction accidents. *Applied Ergonomics*, 36(4): 401–415.

HSE. 2021. *Construction Statistics in Great Britain, 2021. HSE*. [Online]. Available at: https://www.hse.gov.uk/statistics/industry/construction.pdf [Accessed: 24 February 2022].

Lingard, H., Saunders, L., Pirzadeh, P., Blismas, N., Kleiner, B. & Wakefield, R. 2015. The relationship between pre-construction decision-making and effectiveness of risk control. *Engineering, Construction and Architectural Management*, 22(1): 108–124.

Manu, P. 2017a. A model of how features of construction projects influence accident occurrence. In: F. Emuze & J. Smallwood (eds). *Valuing People in Construction*, pp. 41–58. Routledge, Abingdon.

Manu, P. 2017b. Profiling the health and safety risk associated with construction projects at the pre-construction stage. In: F. Emuze & J. Smallwood (eds). *Valuing People in Construction*, pp. 59–77. Routledge, Abingdon.

Manu, P., Ankrah, N., Proverbs, D. & Suresh, S. 2012. Investigating the multi-causal and complex nature of the accident causal influence of construction project features. *Accident Analysis and Prevention*, 48: 126–133.

Manu, P., Ankrah, N., Proverbs, D. & Suresh, S. 2014. The health and safety impact of construction project features. *Engineering, Construction and Architectural Management*, 21(1): 65–93.

Pollack, J. 2007. The changing paradigms of project management. *International Journal of Project Management*, 25(3): 266–274.

Saunders, M., Lewis, P. & Thornhill, A. 2019. *Research Methods for Business Students*. 9th edition. Pearson Education Limited, Harlow.

Suraji, A., Duff, A.R. & Peckitt, S.J. 2001. Development of a causal model of construction accident causation. *Journal of Construction Engineering and Management*, 127(4): 337–344.

U.S. Bureau of Labor Statistics. 2021. *2020 Census of Fatal Occupational Injuries*. [Online]. Available at: https://www.bls.gov/iif/oshwc/cfoi/cftb0340.htm [Accessed: 24 February 2022].

WHO & FAO. 2009. *Risk Characterization of Microbiological Hazards in Food*. Microbiological Risk Assessment Series 17. WHO and FAO, Geneva.

Yin, R. 2003. *Case Study Research: Design and Methods*. Sage Publications, Thousand Oaks, CA.

13 Infrastructure Project Selection and Prioritisation for Socio-Economic Development in Mining Communities of Ghana

A Sequential Mixed Methods Research Approach

Enoch Sackey, Julius Akotia, and Jemima Antwiwaa Ottou

Summary

The planning and execution of infrastructure projects for socio-economic development in the mining regions of Ghana are critical for the inhabitants of those communities. The mining sector generates substantial revenue for the public treasury which leads to national development. However, the mining communities are generally amongst the poorest communities in the country, largely because of a lack of infrastructure development. In the study on which this chapter is based, the subconscious biases that affect decision-makers and mitigating strategies are explored to help draw consensus on the salient project selection and prioritisation (PSP) criteria to facilitate development in communities with skewed infrastructure services and inadequate socio-economic capacity. A sequential exploratory mixed methods research approach was adopted, using semi-structured interviews and a questionnaire survey to collect data from respondents. The findings revealed that infrastructure PSP models to facilitate community development should be focused on two main evaluation measures, namely, prescriptive measures and normative measures. The main contribution of the study in this chapter is that it offers an empirically validated mixed methods research design in which disparate viewpoints were considered to offer objective PSP fidelity. It provides an approach to assess the relative priority of PSP criteria that can be replicated in other communities that are deprived of basic and social infrastructure services.

Introduction

The planning and execution of infrastructure projects for socio-economic development in the mining regions of Ghana are crucial for the indigenous people in those communities and are aligned with the UN Sustainable Development Goals (SDG). Specifically, SDGs 8–10 place emphasis on socio-economic growth and infrastructure development for all by 2030. The implementation of the SDGs lies

DOI: 10.1201/9781003204046-13

mainly with the UN member states, of which Ghana is a signatory. Moldalieva *et al.* (2016) called for signatory countries to focus on the regional and local policy context to achieve the SDGs. Public infrastructure development is capital-intensive and risky to implement and, therefore, requires innovative project selection, prioritisation and financing arrangements.

Thus, the crucial question is whether countries possess the relevant fiscal capacity and budget prioritisation to achieve the SDGs. For instance, many rural communities in Ghana have unmet infrastructure financing needs and various governments in the past have inadvertently been passive in their response to reducing the infrastructure deficit as the resources required are not readily available. Generally, there is a persistent urban–rural disparity in access to basic and social infrastructure services and this has affected the remote mining regions (Booysen, 2003).

Mining imposes socio-economic and environmental costs on host communities, resulting from land acquisition, environmental degradation, pollution, and a high cost of living (Aboka *et al.*, 2018). While mining is an integral part of Ghana's economy and export sector, the mining communities in Ghana are generally poor (Lujala & Narh, 2020). The economic under-performance of natural-resource-rich regions of Ghana is a well-documented phenomenon. Although communities that host mining are entitled to some compensation and mineral royalty transfers, they are still among the poorest communities in Ghana (Dupuy, 2017). Beyond financial resources and other constraints, decision-making pertaining to infrastructure PSP by different stakeholders has been burdened by numerous challenges for these communities. Notable among them are the subconscious biases which are associated with the personal benefits and interests of the various stakeholders involved in the selection and prioritisation of the infrastructure projects for these communities.

The mining sector contributes a great deal of revenues, generated by mineral exports, to the national consolidated funds (CFs) but, underlying the mineral wealth, is the paradox of mining communities generally being amongst the poorest communities in the country, saddled with social, economic and environmental extremes. Therefore, the study in this chapter stemmed from an interest in exploring criteria for the selection and prioritisation of infrastructure projects to optimise scarce resources and enhance socio-economic development in the deprived mining communities in Ghana.

In view of the variations within different interest groups, a sequential exploratory mixed methods research design was adopted for the study to investigate the issues and to determine prioritised criteria for infrastructure project selection, while developing insights for the translation of processes into policy, practice and development outcomes across rural mining communities in Ghana.

In what follows, the legislative efforts made to address the deficits in infrastructure services in the mining communities are discussed. The sequential, exploratory, mixed methods design and its application to advocating infrastructure development are also discussed. Furthermore, the development of a two-phased sequential exploratory mixed method research design, data collection, and analysis is presented. Finally, the contribution of the study to literature about mixed methods is elucidated, identifying implications and lessons learned for using mixed method research design in ways that promote social justice and sustainable development in marginalised communities.

Legislative Policies for Addressing the Deficits in Infrastructure Development in Mining Communities of Ghana

The state of mining communities in Ghana, such as Prestea, Tarkwa, Obuasi, Sefwi-Wiaso, Bibiani and Kenyasi, to mention a few, to this day tells the story of neglect in socio-economic development, as there is ample evidence of a lack of spatial consideration in infrastructure development across many of these communities (Adomako-Kwakye, 2018). Ghana's constitutional provision is that the benefit of revenue earned from mining activities would extend to all Ghanaians, including those in mining communities. However, this has not been the case for the mining communities, as residents have often bemoaned the lack of development in their communities. Ironically, mining exploration displaces many residents, creates unemployment and destroys the environment. Very often the indigenous people in the mining communities lack the capacity to cope with the trend of socio-economic activities associated with mining, thus they are often short-changed and side-lined.

Policy strategies have been proposed to address these multi-faceted issues. Policy strategy that fell within the scope of this study's included the content of local legislation that was intended to increase the share of goods and services supplied domestically (Ackah & Mohammed, 2020) and the Mineral Development Fund Act (MDFA), Act 912 of 2016, that was intended to facilitate socio-economic development in mining communities.

In the MDFA912 (2016), it was mandated that 20% of the mineral royalties, paid by mining companies to the CF through the Ghana Revenue Authority (GRA), must be ceded to the Mineral Development Fund (MDF) to fund development projects in mining communities and some mining institutions. One remarkable feature of Act 912, which was also the core focus of this study, is the establishment of the Mining Community Development Scheme (MCDS) in each mining community. The key role of the MCDS is to spearhead socio-economic development in communities affected by mining operations. According to Section 20(1) of the Act, a Local Management Committee (LMC) in each community would be responsible for administering and overseeing the MCDS. The LMCs are mandated to manage the MCDS by selecting, prioritising and implementing development projects to enhance the livelihood of communities affected by mining activities. The Act requires that 20% of the ceded mineral royalty is disbursed to the MCDS to support socio-economic development of mining and mining-affected communities. In effect, the MDF is intended mainly for development projects in mining communities and for mining institutions.

At the nucleus of a PSP decision model are the large initial investment requirements and the involvement of different stakeholder groups with peculiar needs. The over-arching goal for this study about PSP was to gain insights into how infrastructure PSP criteria can be used to help to close the disproportionate infrastructure gaps in rural mining communities of Ghana and in other diverse communities that face similar disparities in socio-economic development.

Research Methodology and Approach

The application of mixed methods research in the built environment is a natural fit because the built environment and mixed methods share a complementary foundation in their approach to knowledge formation and use. There has been a consistent call in the extant literature for built environment researchers to draw balanced philosophical and methodological inferences (Dainty, 2008; Zou *et al.*, 2014; Harty & Leiringer, 2017) in a bid to propagate methodological diversity and mixed method research study. Dainty (2008), for instance, asserted that methodological pluralism is both legitimate and desirable if established models and understanding are to be questioned and knowledge advanced in the built environment.

In this study, an exploratory sequential mixed methods approach was adopted to explore the issues pertaining to criteria for infrastructure PSP to improve the socio-economic conditions of rural mining communities. The qualitative data were collected first and used to inform the collection of quantitative data in the subsequent phase of the study. The analysis of the qualitative data obtained served as a forerunner to the quantitative research design as it highlighted some vital variables which were worth testing quantitatively (Creswell & Clark, 2017). This enabled the quantitative data to complement the qualitative data, thereby reinforcing the findings of the study.

Data Collection Approach

The study was based on qualitative and quantitative data collected from diverse sources. The qualitative data was collected by the researchers using semi-structured interviews during consultancy engagements with the MDF secretariat to facilitate capacity-building for the LMCs that manage MCDS across the mining regions of Ghana.

Guided by a two-phased sequential exploratory mixed methods design, data collection commenced by investigating decision-making protocols to collect qualitative data from the perspectives of the MDF administrators who manage the mineral royalties for socio-economic development in certain deprived communities of Ghana. The engagement of the researchers with the leadership of the MDF secretariat during the qualitative study yielded 11 ($n = 11$) interviewees from a range of political, administrative and technocratic orientations. The purpose of the qualitative study was to understand how legislative structures influence and promote developments in remote communities. Therefore, the MDF secretariat was selected purposefully for the exploratory component of the inquiry because of its sovereign constitutional mandate to oversee the management of the minerals' royalties.

The semi-structured interview questions were framed to focus more directly on infrastructure development issues within rural mining communities. The semi-structured interviews each lasted between 50–60 minutes and included questions about the role of the MDF secretariat in project initiation and governance, and the legislative requirements for appraising infrastructure projects for funding purposes. The semi-structured interview was structured to explore knowledge, experience and procedures of the interviewees for managing MDF funds. The format of the interview also encouraged the pursuit of new themes and exploration of the contexts of responses with follow-up questions. When no new conceptual themes were emerging, it was assumed that data saturation had been reached (Guest *et al.*, 2006). The interviews were transcribed verbatim and coded using a pattern-matching technique to identify emergent themes.

The responses of the 11 ($n = 11$) participants created the exploratory component of the sequential, exploratory, mixed-methods design. The design of the quantitative research, which succeeded the qualitative study was based on the results of the qualitative study in terms of the funding criteria. Therefore, in the quantitative phase, the role that the mining communities play in decisions regarding infrastructure development and the relative importance of the MDF-sanctioned funding criteria from the perspective of the project beneficiaries was examined.

The quantitative data was subsequently collected from a questionnaire survey of respondents who represented 15 mining communities. A stratified random sampling approach was used to select the target respondents with a view to providing a sample that was considered to be a microcosm of the population. Data was classified into multiple sub-groups based on common socio-demographic interests of the mining communities such as youth, women groups, and traditional rulers. Each stratum was randomly sampled and had an equal chance of being chosen with the intent of finding generalisable results that can be applied to the entire population of the communities (Rahman *et al.*, 2022). A total of 109 survey questionnaires were administered to the LMCs and representatives of mining communities that had shared social and economic concerns within the 15 selected mining communities. The targeted respondents enabled the researchers to garner insights firsthand from representatives with lived experience of the socio-economic challenges confronting their respective communities. A total of 39 responses were received/collated, yielding a response rate of 36%.

The questionnaire survey captured constructs related to experiences in the mining communities, as well as goals, aspirations, and views regarding socio-economic and development issues.

Data Analysis Approach

In the case of the qualitative study, content analysis of the responses was undertaken, and the responses were classified in accordance with the rational, normative indicators purported by the MDF fund managers. The goal of conventional content analysis is to establish more definitive patterns in the data by allowing categories, themes, and insights to emerge directly from the data (Hsieh & Shannon, 2005). The responses were organised into themes informed by the research objectives, according to which legislative policies and the governance criteria of the MCDS were mapped. The researchers reviewed the transcribed data while searching for patterns within the data. Patterns were identified and emergent themes were coded (see Table 13.1) and defined with an emphasis on capturing the decision-making factors related to development funding reported by the participants, as presented in the results section and in Tables 13.1 and 13.2.

The Statistical Package for Social Science (SPSS) was used to conduct statistical analysis of the quantitative component of the data. The data obtained was coded and captured in the SPSS Software for analysis. The statistical analysis comprised descriptive study, mean score analysis, frequencies, percentages, and standard deviation. To determine the criticality/importance of the infrastructure needs, a methodology suggested by Hackman *et al.* (2021) was followed. The relative importance index (RII) and mean score analysis was used to rank the importance of the issues/criteria explored with respondents (Table 13.2). The higher the values of the mean score and RII, the higher the importance that the respondents placed on the criteria. The reason for the use of this statistical analysis was because it promotes generalisation and enables readers to see patterns across studies more easily. The descriptive statistics are summative methods that depict data in a succinct way. The mean is useful in representing the average scores obtained in the study. The infrastructure development gaps that were perceived to impede socio-economic development in mining communities were examined. The rankings of the perceptions of respondents were categorised to make sense of the data. The categorisation and rankings were then synthesised with the main legislation, guidelines, standards and the normative MDF criteria, and were presented in tabular form. The qualitative and quantitative findings complemented each other, achieving scientific cohesion and authentication as the aim of the study was to obtain an overall perception and experience, from both MDF fund managers (MDF officials) and recipients of the funds (LMC representatives), of infrastructure service development. This approach was adopted because socio-economic needs are context-specific and, even within particular communities, there are different stakeholder groups with peculiar needs. Therefore, these issues can be understood better by using a multi-pronged approach involving both quantitative and qualitative methods.

Results

Qualitative Study Findings

The interviewees revealed several factors affecting decisions regarding project funding that either hinged on barriers, such as funding threshold, or were facilitated by feasible strategies for utilising the funds. Determining the most suitable project among alternative community development schemes is discussed as the main goal of the administrators of the royalty from mineral funds.

The analysis of the interviews showed three, main, recurrent criteria when appraising infrastructure projects for funding purposes, comprising (1) effectiveness and impact, (2) project sustainability, and (3) project feasibility. Under the three main criteria, 14 sub-criteria were identified. The effectiveness and impact criteria related to certain characteristics of the project such as resolving identified need (RIN) of the community, being in line with the national development plan (NDP), offering economic opportunity to the community (EOC), positive impact on vulnerable people (IVP), consideration of environmental sustainability (ENS), and balanced portfolio (BPC) in terms of the socio-demographic dispersion of the community. The sub-themes that emerged from the sustainability criteria included the utilisation of local resources (ULR) for the project, assessment of local capacity to ensure continuity (LCC) and operationalisation, assessment of spin-off effect (SOE) and, lastly, the project is expected to complement other factors of production (CFP). Finally, the sub-themes associated with the project feasibility comprised: the practicality of the proposed project (PPP) in terms of technical and financial viability, realistic timeline of the project (RTP) relating to schedule of delivery, availability of skilled workforce (ASW) for construction and operation and, finally, demonstrable experience with regard to the implementation of similar projects (ISP). The criteria and the related sub-criteria are shown in Table 13.1. All projects were evaluated against these criteria to determine which was the most appropriate for achieving value for money.

Table 13.1 MDF criteria for funding mining community development schemes (MCDSs)

Category	*Sub-criteria*	*Code*
Project effectiveness and impact	Address/resolve identified issues/needs within a mining community.	RIN
	Relate to long-to-medium-term national development plan.	NDP
	Offer employment opportunities to local people/ businesses.	EOC
	Impact on vulnerable people (i.e. children, women, disabled and aged) in the community.	IVP
	Environmental sustainability.	ENS
	Balanced portfolio in terms of gender, youth, environment and other identified issues to be addressed.	BPC
Project sustainability	Availability and utilisation of local resources to ensure long-term sustainability (Capex).	ULR
	Development of local capacity to ensure continuity and to serve as a stimulus to aggregate/collective demand.	LCC
	Have spin-off effects (e.g. create local jobs, improve health, improve education, etc.)	SOE
	Serve as a complement to other factors of production.	CFP
Project feasibility	Practicality of the proposed project.	PPP
	Realistic timeline for construction of the proposed project.	RTP
	Availability of skilled workforce during both the construction and operation phases.	ASW
	Demonstrable record of successful implementation of similar projects	ISP

Source: Original.

The rural communities have remained in the doldrums for several years with a huge deficit in social and physical infrastructure. Indeed, the normative funding criteria were initiated by the MDF secretariat to provide nationally-aligned rural infrastructure development that satisfies community dialogue regarding equity, fairness, and economic development in the mining communities. As argued by one MDF official, "the intervention is to break the vicious cycle where local people, especially women and the youth, were excluded from the benefits of mining investments by supporting efforts to better respond to the socio-economic needs of their populations". This was not surprising because one of the main goals of the MDF is the improvement of the living conditions of mining communities and the facilitation of socio-economic development within the framework of equity, fairness, and dignity. The appraisal and funding criteria are specified to ensure that the selected projects will inure to the benefit of the targeted mining communities.

Quantitative Study Results

The mean scores were used to rank the infrastructure development gaps that were perceived to impede socio-economic development in mining communities. For further insight, the descriptive analysis was supported by computing the RII values (as shown below) to obtain rankings and the level of importance of the infrastructure PSP criteria.

The mean scores obtained were compared with the RII value rankings obtained for the criteria in order of importance for the analysis. Computing the RII values made it possible to cross-compare the relative importance of each selection and prioritisation criterion against the mean scores obtained in the descriptive analysis. The higher the mean score and RII value, the higher the level of importance attached to the criteria by respondents. Table 13.2 shows the mean scores, standard deviation, RII and ranks of the perceived infrastructure deficits.

The mean of probability distributions (theoretical mean) for the study was computed by the following formula $\mu\pi = \sum \pi n$, where: $\Sigma\pi$ = the sum of all the means obtained for each criterion; n = total number of criteria. The theoretical mean is defined as the mean of the entire possible sample means obtained (Hackman *et al.*, 2021). All mean scores that were above the theoretical mean were deemed to be critical compared with those that fell below it. Based on the study (Table 13.2), the theoretical mean was found to be 3.13 (Equations 13.1 and 13.2), hence, all the criteria above this range were considered to be critical factors.

The mean of probability distributions (theoretical mean) for the study was computed by the following formula:

$$\mu\pi = \sum \pi n \tag{13.1}$$

where, $\Sigma\pi$ = the sum of all the means obtained for each criteria and n = total number of criteria. Based on the study (Table 13.2), the theoretical mean was determined as

$$\mu\pi = \frac{34.431}{11} \tag{13.2}$$

$$\mu\pi = 3.130$$

Table 13.2 Critical infrastructure needs assessment of local mining communities

Infrastructure deficit	*Min*	*Max*	*Mean score*	*Standard deviation*	*RII*	*Rank*
Access to good quality education and skills development	1	5	3.647	3.299	0.712	3rd
Access to healthcare services	1	5	3.850	3.550	0.770	2nd
Access to good, portable water	1	5	3.870	3.464	0.774	1st
Good transportation network	1	5	2.900	2.757	0.580	7th
Access to electricity/energy	1	5	2.125	1.732	0.425	11th
Affordable housing units	1	5	2.750	2.449	0.550	8th
Community-based recreational facilities	1	5	2.556	2.285	0.511	10th
Community safety and access to security services	1	5	3.200	3.098	0.640	6th
Direct job opportunities (e.g. marketplace and commercial farm)	1	5	3.250	3.010	0.650	5th
Sanitation and drainage system	1	5	2.684	2.555	0.537	9th
ICT and internet services	1	5	3.600	3.347	0.720	4th

Source: Original.

The mean scores for the infrastructure deficits ranged between 3.87 for the highest to 2.125 for the lowest. Out of the 11 ranked gaps in infrastructure services, six of them had their mean values greater than the theoretical mean of 3.13. Access to good portable water was ranked 1st with a mean score of 3.870 and RII of 0.774. Access to healthcare services was ranked 2nd with a mean of 3.850 and RII of 0.770. Access to good quality education and skills development was ranked 3rd with a mean score of 3.647 and RII of 0.712. ICT and internet services were ranked 4th with a mean score of 3.60 and RII of 0.720. Direct job opportunities were ranked 5th with a mean score of 3.250 and RII of 0.537. Community safety and access to security services were ranked 6th with a mean of 3.200 and RII of 0.640. Hence, these six gaps in infrastructure services were deemed to be critical for the mining communities. Since most of the funding for the MCDS for rural infrastructure development was derived from the funds from the royalty for minerals, managed by the MDF secretariat, the identified gaps in infrastructure services must satisfy the imposed criteria (Table 13.1) for accessing MDF funding.

Integration of Qualitative and Quantitative Findings

In this study, the use of a mixed methods research design principally to examine the salient criteria that inform decision-making regarding the selection and funding of development projects to benefit deprived mining communities with disproportionate, basic infrastructural inadequacies and insufficient socio-economic capabilities has been demonstrated. The aim of the qualitative strand was to generate a rich understanding of the rational criteria in determining which development projects best fit strategic goals and, hence, were prioritised to receive adequate funding for implementation. The qualitative data analysis was followed by quantitative data analysis to explain reasons regarding variabilities and relative priority of the criteria from the perspective of broader stakeholder groups that benefited directly from MDF-sanctioned community infrastructure initiatives.

Table 13.1 shows three dominant criteria, comprising (1) effectiveness and impact, (2) project sustainability, and (3) project feasibility, and 14 corresponding sub-criteria identified during the qualitative study which helped management of the royalty for minerals funds to appraise the most viable alternative development projects for financial and administrative support.

Various reasons were offered for the PSP criteria, but they all seemed to be rooted in one core theme: "easy accessibility of basic infrastructure services by indigenous mining communities". For instance, one interviewee (M-F1) stated that:

> Without basic amenities and social infrastructure, any nation's growth and development will be hampered, but the mining communities are plagued by a slew of issues such as unemployment, low productivity, high illiteracy rate, impairing the rural economy and these need to be addressed.

In line with this assertion, another MDF official stated that the selection criterion was intended to "provide a co-ordinated framework for clear-cut strategies to satisfy social dialogue of equity, fairness and economic development in the mining communities".

Therefore, it was clear from the qualitative findings that the funds were geared towards the delivery of crucial community-based infrastructure services to empower and improve the conditions of the mining communities.

Indeed, infrastructure plays an important role in improving economic performance, yet infrastructure projects typically involve substantial up-front capital expenditure (Helm, 2010), and budget constraints in many countries have created formidable obstacles to infrastructure development. Hence, soliciting insights from project beneficiaries in establishing the relative importance of alternative development projects is crucial in making efficient use of constrained resources. This was corroborated by the following statement by a respondent (M-F2): "It is a bottom-up approach. The community themselves would own the projects and the projects belong to the communities. Therefore, they have to identify their needs or whatever project they want to undertake".

From the Mean (M) and RII in Table 13.2, it can be observed that access to potable water, healthcare services, and good quality education and skills development were the top three infrastructure development needs of the rural mining communities (M = 3.87, 3.85, and 3.65, respectively; and RII = 0.774, 0.770 and 0.712, respectively). These findings largely corroborated the results of the qualitative analysis such as the local capacity assessment (LCC) and resolving peculiar health and educational needs of the communities (RIN). The third critical infrastructure need of the communities lent support to the interviewees' NDP selection criteria and complemented the national policy of free basic and senior high education programmes.

The rest of the infrastructure needs which were above the theoretical mean in the quantitative analysis and were deemed to be critical, included ICT infrastructure development (M = 3.6; RII = 0.720). This was followed by job security (M = 3.25; RII = 0.650), and community safety (M = 3.20; RII = 0.640). Job opportunities and community cohesion were among the key drivers for establishing the MDF fund. During the semi-structured interview, one respondent (M-F1) affirmed this by indicating that:

> … rural communities are rife with seasonal unemployment amongst smallholder farm workers characterised by vulnerable workers and low productivity. Worse still, youths in mining communities face unemployability challenges. These issues informed the establishment of the MDF fund.

Therefore, the results revealed that what the fund managers believed might be the prioritised needs of the project beneficiaries might differ from the relative priorities of the beneficiaries.

The implication of this finding is that there is a risk of the PSP process losing its relevance because of perceived biases associated with imbalanced stakeholder representation during dialogues. Two divergent opinions between the management of MDF funds with their rational notions, and project beneficiaries with lived experiences and peculiar needs, might result in the scheme not achieving the needed benefits. The inimitable requests of community members should drive the infrastructure development choices. The MDF scheme should be able to support a wide range of social interventions, while supporting the target interest groups to locate the resources required to initiate developments that align with national interests and local priorities. Further, as a policy, the MDF should make provisions to mitigate unfair biases while preventing a perceived lack of agency in the process to ensure buy-in and commitment to the cause.

Implication of the Use of Mixed Method Research Design

The study contributes to the current discourse on a mixed methods approach to advance the understanding of PSP for the built environment in three ways. First, it broadens the conceptualisation of an infrastructure PSP model, which is critical for closing the disparity in access to development across certain deprived communities. It is also critical for balancing essential voices and power in the deliberation. This stance helps to avoid subconscious biases and one-sidedness by blending the lived experience of project beneficiaries with legislative decisions of authorities through mutual reflection towards achieving consensus.

Second, the sequential exploratory mixed methods approach helps to explore the congruence and/or divergence of perceptions of context, interventions and outcomes. The convergence of the mixed data offered breadth and depth to the infrastructure selection and prioritisation process. The qualitative data provided a view of funding criteria from the perspectives of legislation and governance, and the quantitative data indicated the relative importance of the criteria with a sharper sense of context regarding the selection and development of infrastructure projects.

Third, this process can be transferred to other contexts to help to establish consensus on the salient criteria and their relevance to the whole process of selecting and prioritising projects to benefit communities with disproportionate infrastructure deficits and inadequate socio-economic prowess.

Conclusions

In this study, a theoretical balance was struck between normative and prescriptive viewpoints concerning decision-making by different interest groups regarding infrastructure project selection, funding, and development. For this purpose, descriptive analytic techniques were employed to analyse the quantitative data collected. Infrastructure PSP frameworks could be informed by the dichotomy between descriptive and normative theories. On one hand, the contextual antecedents and real, human situational context are considered in descriptive decision theory and, on the other hand, a decision in normative theory is regarded as a rational selection problem, thereby defining how a rational objectivist should deduce the logically best alternative project.

The study in this chapter lays the foundation for further studies on the topic with data that stretch over broader mining communities for the analysis of gaps and opportunities in infrastructure development for alternative funding arrangements to augment the MDF funding initiative. The synthesis of mixed methods research was focused on the infrastructure selection and prioritisation (PSP) phase, but it is noted that the implementation phase can equally hinder the realisation of the expected socio-economic benefits. Indeed, future research would benefit from assessing the subsequent phase of infrastructure deployment and the realisation of post-implementation benefits from the MDF funds as perceived by the targeted beneficiary mining communities.

References

Aboka, E.Y., Cobbina, D.J. & Dzigbodi, D.A. 2018. Review of environmental and health impacts of mining in Ghana. *Journal of Health and Pollution*, 43: 45–49.

Ackah, C.G. & Mohammed, A.S. 2020. Local content law and practice: The case of Ghana. In: J. Page & F. Tarp (eds). *Mining for Change*, pp. 139–160. Oxford University Press, Oxford.

Adomako-Kwakye, C. 2018. Neglect of mining areas in Ghana: The case for equitable distribution of resource revenue. *Commonwealth Law Bulletin*, 44(4): 637–651.

Booysen, R. 2003. Urban and rural inequalities in health care delivery in South Africa. *Development Southern Africa*, 20(5): 659–673.

Creswell, J.W. & Clark, V.L.P. 2017. *Designing and Conducting Mixed Methods Research*. 3rd edition. Sage Publications, Thousand Oaks, CA.

Dainty, A. 2008. Methodological pluralism in construction management research. In: A. Knight & L. Ruddock (eds). *Advance Research Method in Built Environment*, pp. 1–12. Blackwell Publishing Ltd, Oxford.

Dupuy, K. 2017. Corruption and elite capture of mining community development funds in Ghana and Sierra Leone. In: A. Williams & P. Le Billon (eds). *Corruption, Natural*

Resources and Development: From Resource Curse to Political Ecology, pp. 69–79. Edward Elgar Publishing Limited, Cheltenham.

Guest, G., Bunce, A. & Johnson, L. 2006. How many interviews are enough? An experiment with data saturation and variability. *Field Methods*, 18(1): 59–82.

Hackman, J.K., Ayarkwa, J., Osei-Asibey, D., Acheampong, A. & Nkrumah, P.A. 2021. Bureaucratic factors impeding the delivery of infrastructure at the Metropolitan Municipal and District Assemblies (MMDAs) in Ghana. *World Journal of Engineering and Technology*, 9(3): 482–502.

Harty, C. & Leiringer, R. 2017. The futures of construction management research. *Construction Management and Economics*, 35(7): 392–403.

Helm, D. 2010. Infrastructure and infrastructure finance: The role of the government and the private sector in the current world. *EIB Papers*, 15(2): 8–27.

Hsieh, H. & Shannon, S. 2005. Three approaches to qualitative content analysis. *Qualitative Health Res*earch, 15(9): 1277–1288.

Lujala, P. & Narh, J. 2020. Ghana's minerals development fund act: Addressing the needs of mining communities. *Journal of Energy & Natural Resources Law*, 38(2): 183–200.

MDF Act. 2016. *Minerals Development Fund Act, 2016 (Act 912).* The Parliament of the Republic of Ghana, Accra.

MDF Board. 2020. *Guidelines for the Disbursement of Funds to Local Management Committees for the Management of Mining Community Development Scheme*. The MDF Secretariat, Accra.

Moldalieva, J., Muttaqien, A., Muzyamba, C., Osei, D., Stoykova, E. & Le, N. 2016. *Millennium Development Goals (MDGs): Did They Change Social Reality?* United Nations University-Maastricht Economic and Social Research Institute on Innovation and Technology (MERIT), Maastricht.

Morgan, D.L. 2007. Paradigms lost and pragmatism regained: Methodological implications of combining qualitative and quantitative methods. *Journal of Mixed Methods Research*, 1(1): 48–76.

Peterson, M. 2017. *An Introduction to Decision Theory*. Cambridge University Press, Cambridge, New York, NY and Port Melbourne.

Rahman, M.M., Tabash, M.I., Salamzadeh, A., Abduli, S. & Rahaman, M.S. 2022. Sampling techniques (probability) for quantitative social science researchers: A conceptual guideline with examples. *Seeu Review*, 17(2): 42–51.

Zou, P.X., Sunindijo, R.Y. & Dainty, A.R. 2014. A mixed methods research design for bridging the gap between research and practice in construction safety. *Safety Science*, *70*, 316–326.

14 Exploratory Sequential Mixed Method Research to Investigate Factors Affecting the Reputation of PFI/PF2 Projects in the UK

Stanley Njuangang, Henry Abanda, Champika Liyanage, and Chris Pye

Summary

The UK Government introduced the private finance initiative (PFI) as a measure to work closely with the private sector in the provision of public projects. Despite its popularity in the 1990s, it presently faces strong criticisms for failing to deliver value for money. Different measures, that is Private Finance 2 (PF2), by successive UK Governments have not changed the situation. The exploratory sequential mixed methods research design applied to investigating the reputation of PFI/PF2 projects in the UK is presented in this research. Grounded theory was the primary method for conducting the qualitative research phase. Thereafter, the themes were distilled and constituted the basis of a questionnaire survey using a five-point Likert Scale. The questionnaire was despatched to selected stakeholders in the construction industry and the data was analysed using the Relative Importance Index (RII) and Kruskal Wallis Test. Using the results of the RII, the most important factors affecting the reputation of PFI projects were selected for the development of a theoretical framework. Despite what appears to be disagreements amongst researchers on the true value of mixed methods research, the findings from using this approach made it possible for conclusions to be drawn from multiple perspectives. Hopefully, issues regarding the paradigmatic position of mixed methods research will be overlooked in favour of its contribution to investigating contemporary issues.

Introduction

The PFI scheme remains the most used type of public–private partnerships (PPP) in the UK's National Health Service (NHS). The PFI involves an arrangement whereby a private consortium, that is a special purpose vehicle (SPV), may be asked to build and operate a new or redeveloped facility for a period of up to 30 years. PFI provides the government with an opportunity to widen the scope of private sector involvement in the provision of infrastructure and public services (Mercer & Whitefield, 2018). In the healthcare sector, it was introduced to

DOI: 10.1201/9781003204046-14

"... attract private sector finance, management skills and expertise into the provision of public sector facilities and services" (Akintoye *et al.*, 1998: 9).

Despite the government's rhetoric about the potential benefit of the scheme, there was the aspect that the private sector was very reluctant to embrace the idea entirely. As a result, in 1994, the new Chancellor, Kenneth Clark, made it mandatory for all capital projects, requiring the approval of the treasury, to explore the option of PFI in their proposals. According to Broadbent *et al.* (2004), such a "universal testing" policy was adopted without due consideration of the cost involved (especially in terms of legal and financial advisory costs) or exploration of the true value of the newly introduced change. These reasons explain why the policy was abandoned by the Labour Government after they won the 1997 elections. The Labour Government regarded the time and money spent on trying to develop models for "universal testing" as being a waste of valuable resources.

Subsequent changes instituted by the Labour Government did not immediately stimulate PFI schemes in the healthcare sector (Broadbent & Gill, 2003). According to Patel and Robinson (2010), it took ten years after the launch of the first wave of PFI hospitals in 1995 for the NHS to witness any significant amount of PFI activities. By the onset of the last financial crises of 2007/08, there were 728 PFI projects, with a capital value (total worth of infrastructure assets) of just over £56 billion (Booth & Starodubtseva, 2015). However, following the crises, the cost of private borrowing increased and parliament became critical of PFI schemes for failing to demonstrate value for money to the taxpayers (NAO, 2018).

In 2012, the coalition government introduced an amended form of the PFI scheme, called PF2. Although PF2 maintained many of the features of the original PFI scheme (Mercer & Whitefield, 2018), it was supposed to improve transparency and enforce greater accountability in the procurement and operation of projects. Also, in the new scheme, the coalition government held a minority equity stake and reduced the provision of soft services, that is cleaning waste management by SPV. Even the creation of the Infrastructure and Projects Authority (IPA) to provide support functions to government departments for infrastructure and projects met with some criticisms. The National Audit Office (NAO) (2020), in its review, found that NHS Trusts were strategically unprepared to assume responsibility over PFI projects at the end of contract terms.

Following the criticisms levelled against the PFI scheme, in 2018, the UK Government announced that it was going to disband its application. Therefore, the fundamental issue for this research study was to investigate factors that affect the reputation of PFI/PF2 projects in the UK. Past and present literature was examined in depth to identify reasons for the unpopularity of PFI/PF2 schemes in the UK Healthcare Sector. Previous research in this area was focused mostly on measures to improve the financial performance of PPP/PFI schemes. Li *et al.* (2005) identified effective procurement, favourable economic conditions, and available financial market as being key issues in the implementation of PPP/PFI projects. Conversely, Carrillo *et al.* (2008) conducted their research on the participation, barriers, and opportunities in PFI in the UK.

An Overview of PFI/PF2 Projects

In a typical PFI project, the public and private sectors enter into a long-term contractual arrangement for up to 40 years. The SPV is formed specifically for a single project and usually comprises a construction company, a facilities management (FM) function and a financier. In a typical PFI arrangement, 90% of the finance is drawn from debt and 10% from equity (National Audit Office, 2010). Since the debt portion of the financing consists of bank loans and/or bonds, the SPV pays interest for the risk incurred. The supposed interest is usually divided into two, that is the interbank rate, reflecting the general market risk, and the loan margin, reflecting the project-specific risk. As risk stabilises over time, the variable bank rates are replaced with fixed monthly payments. The variable rate is converted into a fixed, long-term, interest rate that is paid over the life of the project.

Although the introduction of the PFI scheme might have attracted private finance and resulted in the building of new public projects, there is no clear evidence that it was better than other forms of procurement. According to the UK Government, "it only uses PFI where appropriate, and where it can deliver 'value for money' benefit" (Hill & Collins, 2004). Both Conservative and Labour Governments have rebuffed the idea that PFI was introduced to secure "off balance sheet" treatment for some public projects. According to the government, the public sector lacked the sort of expertise, innovation, and project management skills required for the successful delivery of projects. As a result, most public projects that were delivered through the traditional route, incurred time and cost overruns. Therefore, through PFI schemes, the government hoped to achieve better value for money for the taxpayers. With over 88% of PFI projects being completed on time and within budget, proponents of the PFI scheme claimed that it offered value for money to taxpayers (Roe & Craig, 2004). According to the NAO, the vast majority of PFI projects were constructed close to the required time frames and budget (House of Commons, 2015).

So far in the UK, there have been success stories since the adoption of the PFI scheme. Unlike in the healthcare sector, with complex projects, most of the PFI success stories were in transport and prison services. For example, HMP Rye Hill was built in 16 months; following the traditional procurement route, the same prison would have taken three years to construct (Roe & Craig, 2004). In the transport sector, the widening of the M40 was also an example of a PFI project that was delivered early. The suggestion that these projects were completed on time does not mean that they performed well during the procurement, maintenance and FM phases of the projects. The Mid Yorkshire Hospitals Trust (MYHT) was an example of a £311 million PFI hospital that experienced problems during its operational phase. Faced with inefficiency and tightening of government spending on public services, it became difficult for MYHT to maintain monthly unitary payments (UNISON, 2013). In the transport sector, the government has bailed out PFI/PPP projects such as the Channel Tunnel Rail Link and the Royal Armouries Museum.

From the foregoing discussion, it is evident that the story of the success of PFI projects is a mixed one. In a separate report to HMP, the NAO (2003), cited by

Edwards *et al.* (2004: 20), noted that "the use of the PFI is neither a guarantee of success nor the cause of inevitable failure. Like other forms of providing public services, there are successes and failures …". One thing is certain, even with the abandonment of new PFI schemes, the UK Government will still need the participation of the private sector to finance its infrastructure projects that currently amount to between 1.0% and 1.2% of the GDP each year. With the last PFI project set to end by 2050, relevant stakeholders must seek to manage its performance successfully. In 2016–2017 alone, the total unitary charge payment made by the health bodies was £2 billion, representing 1.7% of the total cash budget for the Department of Health and Social Care (NAO, 2018). In the next section, the focus is on discussing the research methodology used to identify factors that affect the reputation of PFI/PF2 projects in the UK.

Research Methodology

As shown in Figure 14.1, exploratory sequential mixed methods research was used to investigate factors that affect the reputation of PFI/PF2 projects in the UK. According to Teddlie and Tashakkori (2008), this approach involves the analysis of qualitative data with the aim of developing a quantitative instrument to explore the research problem further. Although exploratory sequential mixed methods research has been applied differently by many authors, its primary purpose is to identify variables for use in quantitative measurement (Edmonds & Kennedy, 2017). This research was consistent with this purpose, where grounded theory (GT) was applied to identify and categorise factors that affect the reputation of PFI projects in the UK. Using the findings of qualitative research in the development of a quantitative research instrument increases the validity and reliability of the results (Shiyanbola *et al.*, 2021).

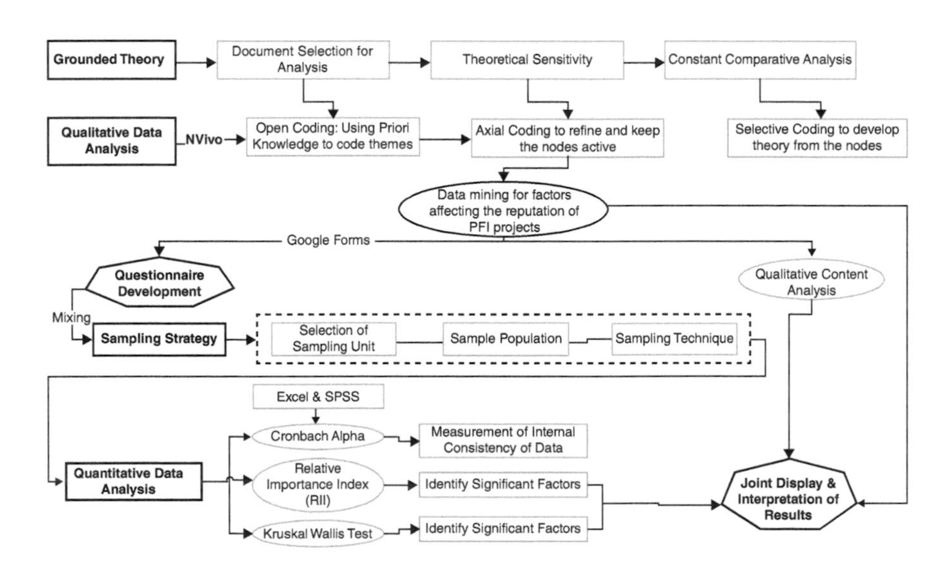

Figure 14.1 Exploratory sequential mixed methods for identifying factors affecting the reputation of PFI/PF2 projects

Although results based on the qualitative data are "exploratory" in nature (Creswell & Clark, 2018), they provide useful information for the interpretation of the overall results of the research. The results obtained from the qualitative content analysis are shown in the sub-section about GT, together with the Pearson Correlation Coefficient, which shows mainly the similarity of the material coded in the different nodes. The primary purpose of conducting these analyses was to gain preliminary knowledge to inform further investigation. In the quantitative phase, data was analysed using Cronbach's alpha, RII, and the Kruskal Wallis Test. Of importance is the level of integration between the qualitative and quantitative research methods. In this case, the point of interface between the qualitative and quantitative research phases was in the development of the research instrument for conducting the quantitative research and development of the theoretical framework. The list below shows the different stages in carrying out the qualitative and quantitative phases of the research:

- **Qualitative research phase:**
  - Grounded theory and selection of relevant research material:
    - Open coding: developing the themes
    - Axial coding: developing the parent and child nodes
    - Development of the research instrument
- Selective coding: development of the theoretical framework
- **Quantitative research phase**
  - Demographic analysis
  - Quantitative data analysis and results
- **Joint display of the qualitative and quantitative results**

Qualitative Research Phase

In the current study, qualitative research was conducted using the GT method, involving open, axial, and selective coding of themes in selected research materials. The process of coding the material into the different "parent" and "child" nodes was done using the application of qualitative QSR NVivo 12 Software. The themes coded in the nodes were refined and analysed using exploratory content analysis. These processes are explained in-depth in the following sub-sections.

Grounded Theory and Selection of Research Materials

Developed by Glaser and Straus in 1967, GT is regarded by many as the metaphor of qualitative research. This is because GT allows for critical thinking and discovery without prior knowledge (Mengye & Simon, 2021). Irrespective of the strand of GT, there are three fundamental principles for the development of the theory (Bulawa, 2014). The first, which is called theoretical sensitivity, enables the researcher(s) to become deeply immersed in relevant research material to understand salient issues about the research topic. On the other hand, theoretical sampling allows for the coding of data into different categories and sub-categories for the development of theory. Based on constant comparison, as new material emerges, the researcher moves forwards and backwards, refining the codes for sufficiency.

In similar research, in which literature was analysed using GT, Wolfswinkel *et al.* (2011) set the inclusion/exclusion criteria for the research material. For example, in the current research, all material for conducting GT analysis has relevance to PFI/PF2 projects in the UK. Databases, from which research materials were obtained, included Science Direct, Emerald, and Google Scholar. Some of the research materials were also drawn from relevant government bodies and online sources. The process of sourcing the research material for coding started with basic searches using words such as "PFI", "procurement", "reputation", and "PPP". This produced a list of materials about PFI but did not reveal sufficient information about factors that affect the reputation of PFI projects in the UK. As a result, the search criteria were broadened to include words such as "criticism of PFI projects in the UK", "growth of PFI in the UK" and "popularity of PFI in the UK". Even then, the materials that were produced were not directly related to the research topic. Nonetheless, the search resulted in the identification of additional material from which the relevant issues could be inferred.

In total, 39 documents were selected for the next phase of the research that involved coding using QSR NVivo. Out of this number (see Table 14.1), nine documents were drawn from journal papers (represented by green icons), 13 from UK Government Departments (red icons), and 14 from other professional organisations (purple icons). Only 25 of these documents contained material that was directly related to factors that affect the reputation of PFI projects in the UK. Information contained in the remaining 14 documents was used to support key arguments and facts in the literature review process. How these documents were analysed using the Straussian GT – open, axial and selective coding is shown in the following sub-section.

- Coding and Development of the Research Instrument

Strauss and Corbin (1990: 61) defined open coding as "… the process of breaking down, examining, comparing, conceptualizing, and categorizing data". In the present research, open coding was used primarily to identify and code materials for the development of categories and sub-categories. The process was assisted by "broad-brush" coding using qualitative QSR NVivo Version 13 Software. The software was selected because it allowed for the creation of "parent" and "child" nodes. Figure 14.2 shows the coding of materials from Carrillo *et al.* (2008) and European Services Strategy Unit (2017) into the "parent" node, called "procurement".

With regards to the factors that affect the uptake of PFI projects in the UK, the material was coded initially in seven parent nodes, namely, administration of contracts, controversies, financial factors, operation, risk allocation, staffing issues and sustainability. Interrogation of the nodes suggested the need for refinement and re-categorisation of parent nodes. In other areas, nodes were created, that is "definition of PFI", "benefits and merits of PFI", "PFI cases with issues" to

Table 14.1 Selected documents with coding references and nodes

Name	*Assigned codes*	*Codes*	*References*	*Modified on*	*Modified by*
Bing Li et al. (2005)	JCP-1	3	6	16-06-2020 03:18	SN
Carrillo et al. (2008)	JCP-2	8	12	16-06-2020 03:19	SN
Hill and Collins (2004)	JCP-3	4	11	16-06-2020 03:28	SN
Jaw-Kai Wang	JCP-4	0	0	16-06-2020 03:23	SN
Neufville et al. (2007)	JCP-5	0	0	16-06-2020 03:30	SN
Parker (2009)	JCP-6	0	0	16-06-2020 03:31	SN
Akbiyikli and Eaton (2004)	JCP-7	1	8	16-06-2020 03:18	SN
Gao et al. (n.d)	JCP-8	1	1	16-06-2020 03:26	SN
Ning et al. (2004)	JCP-9	0	0	16-06-2020 03:35	SN
House of Commons (2008)	GOP-1	8	172	24-06-2020 18:58	SN
House of Commons (2003)	GOP-2	3	33	24-06-2020 17:29	SN
House of Commons (2011)	GOP-3	7	23	24-06-2020 17:29	SN
National Audit Office (2008)	GOP-4	2	6	24-06-2020 17:29	SN
House of Commons (2011b)	GOP-5	1	1	24-06-2020 17:29	SN
Northern Ireland Assembly (2011)	GOP-6	1	4	24-06-2020 17:30	SN
National Audit Office (2010)	GOP-7	6	10	24-06-2020 17:30	SN
National Audit Office (2009)	GOP-8	6	15	24-06-2020 17:34	SN
HM Treasury (2003)	GOP-9	1	1	24-06-2020 17:34	SN
House of Commons (2015)	GOP-10	8	24	24-06-2020 17:43	SN
HM Treasurey (2016)	GOP-11	0	0	24-06-2020 17:44	SN
National Audit Office (2018)	GOP-12	12	52	24-06-2020 17:44	SN
House of Commons (2003a)	GOP-13	0	0	24-06-2020 17:44	SN
McKee et al. (2006) – WHO	PRD-1	8	23	24-06-2020 18:57	SN
Roe and Craig (2004) -CPS	PRD-2	10	52	24-06-2020 17:33	SN
Pollock et al. (2013)	PRD-3	3	4	24-06-2020 19:05	SN
Hellowell (2014) – CHPI	PRD-4	3	6	24-06-2020 17:35	SN
DeadWeight – Unison	PRD-5	3	4	24-06-2020 19:05	SN
Edwards et al. (2004) – ACCA	PRD-6	5	13	24-06-2020 19:03	SN
Lister (2003) – UNISON	PRD-7	2	4	24-06-2020 17:35	SN
European Services Strategy Unit (2017)	PRD-8	6	10	24-06-2020 17:42	SN
NHS Support Federation (n.d)	PRD-9	4	5	05-05-2020 23:40	SN
PFIreport	PRD-10	0	0	05-05-2020 23:40	SN
Broadbent et al. (2004) – CIMA	PRD-11	0	0	24-06-2020 19:04	SN
PricewaterhouseCoopers	PRD-12	0	0	24-06-2020 19:08	SN
CHPI-PFI-Return-Nov14-2 (2)	PRD-13	0	0	24-06-2020 19:05	SN
Walker and Tizard (2018) – Smith Institute	PRD-14	0	0	24-06-2020 17:44	SN

Source: Original.

Name: Procurement

<Files\\PFI\\Carrillo et al. (2008)> - § 1 reference coded [0.47% Coverage]

Reference 1 - 0.47% Coverage

The figures show that the majority of client and construction organizations believe that the bidding costs of PFI are higher, regardless of project size, whereas only a third believe these costs are higher for the design and construction phases. The main reason for this is the cost of the specialist expertise required during the bidding stage and the lengthy negotiation periods for PFI projects.

<Files\\PFI\\European Services Strategy Unit (2017)> - § 5 references coded [0.41% Coverage]

Reference 1 - 0.05% Coverage

Abandoned projects
Sixteen PFI/PPP projects were cancelled at a cost of £114.3m. The cost would be significantly higher if information was publicly available for other cancelled projects.

Reference 2 - 0.14% Coverage

A large and complex contract is at the centre of every PFI/PPP project. A standard draft contract is amended and developed as procurement proceeds up to the point of financial closure. The final contract or project agreement can range from a few hundred to several thousand pages. But no matter how comprehensive they are, virtually all contracts are incomplete in practice (Hart, 2003), because they cannot predict future events and changing economic and social needs. Tirole (1999) identifies three reasons for incomplete contracts:

Figure 14.2 Sample of themes coded in node, called "procurement"

help with the literature review process. Figure 14.2 shows a view of the material and percentage that was coded from two different sources (Carrillo *et al.*, 2008 – 0.47%; European Services Strategy Unit, 2017 – 0.41%) in the parent node called "procurement".

The process of axial coding involved working through the nodes looking for similarities, relationships and opposites of factors that affect the reputation of PFI projects. Through constant comparison, that is moving forwards and backwards to update and keep the nodes active, the decision was taken to split some of the nodes. For example, the child node called "administration of PFI contracts" was cascaded under the parent node called "Procurement". Similarly, re-financing was added as a child node under the node called "Finance". So far, the factors that affect the reputation of PFI projects in the UK were grouped under seven parent nodes, that is procurement, administration, finance, operation, staffing, risk, and sustainability (Table 14.2 shows a list of factors under "Procurement").

Whilst QSR NVivo Software is used mainly for qualitative data analysis, it nonetheless provides a function for cluster analysis, using similar words and attributes coded in the different parent and child nodes. Given the subjectivity of the results, the use of clusters in this research, and those of the qualitative content analysis, was mainly exploratory. The results of the Pearson Correlation Coefficient showed correlation between "finance and administration of PFI contracts" ($p = 0.768$); "staffing issues" and "procurement" ($p = 0.707$) (refer to Table 14.3).

Table 14.2 Refined factors in the procurement nodes

Primary nodes		*Factors affecting uptake of PFI projects in the UK* (n = 23)	*References (x)* (n = *31)*							
			PR-1	*PR-2*	*PR-3*	*PR-4*	*PR-5*	*PR-6*	*PR-7*	*%*
Procurement	1	High bidding prices regardless of project size	X							14
	2	Abandonment & cost terminating of PFI projects		X					X	29
	3	Lack of public information about cost of project		X						14
	4	Over specification & complexity of projects that are difficult to understand thus deterring potential bidders		X	X		X			43
	5	Difficulty predicting demands, changes in economic & social needs		X						14
	6	Revision of public policy priorities, technologies and operations		X						14
	7	Performance of private sector consortia (construction, banks, facilities management contractors)		X						14
	8	Emphasis on achieving project outcomes rather than quality of inputs, processes & outputs		X						14
	9	Inability to resolve & enforce terms of contract without courts		X						14
	10	Cost on specialist expertise (i.e. legal fee) to advice on contract	X						X	29
	11	Lengthy contract negotiation periods	X		X					29
	12	Incentives that are unconnected to value for money & flexible designs			X	X				29
	13	Weak assessment criteria resulting to unsuitable PFI projects			X		X			29
	14	Failure to compare PFI projects against government borrowing			X					14
	15	Increased cost for PFI projects at the stage of financial close			X					14
	16	Run-down public-sector departments, that is architecture & planning to handle procurement				X				14
	17	Public sector inability to provide complete drawings & specifications of end product				X			X	29
	18	Failure to incorporate design features that benefit users of public buildings							X	14
	19	Imbalance of power, skills & knowledge between public and private sectors				X				14
	20	Unclear strategic reasons for using PFI					X			14
	21	Error ridden financial modelling that favour PFI						X		14
	22	Lack of system to collect comparable data on similar PFI projects						X		14
	23	Delays caused by 'log jams', start-stops approach by Central Government							X	14
		Total references	9.6	26	19	13	9.6	9.6	13	100

Source: Original.

Table 14.3 Correlation between the different nodes (QSR NVivo Version 12)

Code A	*Code B*	*Pearson correlation coefficient*
Financial factors	Administration contracts	0.768
Staffing issues	Administration contracts\procurement	0.707
Administration contracts\procurement	Administration contracts	0.691
Administration contracts\procurement	Financial factors	0.530
Risk allocation	Administration contracts	0.477
Staffing issues	Administration contracts	0.397
Risk allocation	Administration contracts\procurement	0.313
Operation	Administration contracts	0.250
Risk allocation	Operation	0.250
Risk allocation	Financial factors	0.233
Staffing issues	Risk allocation	0.215
Staffing issues	Financial factors	0.211
Administration contracts\procurement	Operation	0.195
Administration contracts\procurement	Controversies	0.182
Financial factors\re-financing	Financial factors	0.171
Staffing issues	Operation	0.147
Operation	Financial factors	0.133
Sustainability	Staffing issues	0.099
Operation	Controversies	0.060
Sustainability	Controversies	0.041
Sustainability	Financial factors\re-financing	-0.066
Staffing issues	Controversies	-0.088
Financial factors\re-financing	Operation	-0.098
Controversies	Administration contracts	-0.137
Risk allocation	Controversies	-0.137
Financial factors\re-financing	Administration contracts\procurement	-0.141
Sustainability	Operation	-0.142
Staffing issues	Financial factors\re-financing	-0.156
Financial factors\re-financing	Controversies	-0.187
Financial factors	Controversies	-0.195
Financial factors\re-financing	Administration contracts	-0.204
Risk allocation	Financial factors\re-financing	-0.204
Sustainability	Administration contracts\procurement	-0.204
Sustainability	Administration contracts	-0.295
Sustainability	Risk allocation	-0.295
Sustainability	Financial factors	-0.385

Source: Original.

With the help of qualitative content analysis, raw data in the different nodes were distilled and consolidated as variables in the questionnaire. In total, 91 factors were identified and categorised into the initial parent nodes as follows: procurement (23 factors), administration (15 factors), finance (23 factors), operations (seven factors), staffing (six factors), risk (nine factors), and sustainability (8 factors). Table 14.2 shows a list of documents (PR1, PR2, PR3, PR4, and PR5) from which factors related to procurement were taken. The table also shows the number of times a factor was coded from each of these documents.

Unlike in open and axial coding, the material used in the selective coding was refined, that is 91 factors that affect the reputation of PFI projects. Following interrogation, the lists of factors in the different categories were integrated into a central concept for the development of theory. In this research, the core category or central concept was the category or node called "procurement of PFI projects". Given the number of factors involved, they were analysed for level of significance, using the RII. In terms of selective coding, the refined list of factors and results of RII were used in the development of a theoretical framework of factors that affect the reputation of PFI/PF2 projects in the UK (see Figure 14.2).

Quantitative Research Phases

Quantitative research involves subjecting quantitative data to rigorous quantitative analysis in a formal or rigid fashion (Goddard & Melville, 2001). Quantitative research can be sub-divided into simulation, experimental and inferential methods of research (Goddard & Melville, 2001). Experimental research involves research in which the researcher has greater control over the research environment. The simulation approach is appropriate for researchers interested in building models for the understanding of future conditions. In this study, inferential research was applied based on a survey. This involved conducting an in-depth study of a sample of the population (through questioning), with the intention of inferring the characteristics of the rest of the population.

With the use of questionnaires, enabled by Google Forms, the 91 factors that were found to affect the reputation of PFI projects were presented in a five-point Likert Scale. Google Forms made it easier for the questionnaire to be circulated electronically to a wider audience. Out of the 250 questionnaires distributed through LinkedIn, and Facebook groups for construction professionals, that is quantity surveyors, construction project managers, building surveyors, etc., only 26 were returned successfully. Without being able to increase the number of responses, the decision was taken to analyse the data using a non-parametric test. According to Pett (2016), a non-parametric test could accommodate small sample sizes and data with irregular sample distributions. It is not uncommon for mixed methods research to have varied response rates (Abowitz & Toole, 2010). Most of the respondents (56%) had working experience of 11+ years, working in the construction industry. They were mainly quantity surveyors, project managers, building surveyors (working in different areas of the construction industry), and academics. In the next section, the quantitative data analysis and results are discussed.

Quantitative Data Analysis and Results

Data was analysed using quantitative Microsoft Excel and SPSS Software. While the former was used to analyse the RII of factors that affect the reputation of PFI projects, the latter was used to examine the level of statistical differences between the different sub-groups. According to Johnson and LeBreton (2004, cited by Somiah *et al.*, 2015: 120), "RII aids in finding the contribution a particular variable makes to the prediction of a criterion variable both by itself and in combination with other predictor variables". The results for RII were in the range $0 \leq \text{RII} \leq 1$, with higher scores indicating higher levels of relative importance. The formula below was used to determine the RII:

$$\text{RII} = \frac{\Sigma \text{W}}{(\text{A} \times \text{N})} \tag{14.1}$$

where W = weighting given to each statement by the respondents and ranges from 1 to 5; A = higher response; integer (5); and N = total number of respondents.

Table 14.4 shows the results of the RII for some of the factors in the financial category. Out of the 91 factors, the failure of government departments to recognise the complexity of refinancing emerged as the 5th most important factor affecting the reputation of PFI projects. The Kruskal Wallis Test was used to establish the level of disagreement between the different groups of respondents according to the type of organisation, experience, and profession (see Tables 14.4 and 14.5). The test could be used to compare the mean of three or more distinct groups. In the current research, under organisation, there were four groups – client, main contractor, subcontractor, consultancy, and academia. Conversely, under profession, there were six groups – director, building surveyor, project manager, quantity surveyor, and academics. The test made it possible to establish whether significant differences existed in the way the different groups rated factors that affect the reputation of PFI projects. *Post hoc* analysis involved using the mean scores to indicate groups with higher and lower levels of consensus. The level of statistical significance for this research was set at $p = 0.05$.

Given the scope of this research, it was not possible to discuss all the results of the quantitative analysis. However, under the category called "administration of PFI contracts", there were significant differences ($p = 0.027$) between the different groups of professionals regarding "lack of public scrutiny of PFI on grounds of commercial confidentiality". Disagreement on this issue was mainly between academics (mean = 4.30) and QS (mean = 2.50), and PM (mean = 2.50). On the issue of "lack of information sharing and transparency", there was a lack of consensus (p = 0.030) between the different groups of respondents. On this factor, there was a stronger level of agreement between academics (mean = 4.29) and QS (mean = 4.25);there was a lower mean score for project managers (mean = 2.3) and directors (mean = 3.00).

Joint Display of the Qualitative and Quantitative Results

The results of this research include those of the qualitative and quantitative data analysis. The key results of the qualitative data analysis include those for the identification of factors that affect the reputation of PFI projects. Table 14.4 shows how

Table 14.4 Financial factors that affect the reputation of PFI projects

		Likert scales								*Kruskal Wallis test*		
		1	*2*	*3*	*4*	*5*	*RII*	*Ranking for category*	*Overall ranking*	*Experience*	*Organisation*	*Profession*
	Financial issues (*N* = 23)						*0.681 (4)*					
FI-1	High cost involved in the tendering process		2	6	8	9	0.762	3	9	0.793	0.139	0.390
FI-2	Cost constraints initiating innovation in PFI projects		3	6	11	4	0.677	11	46	0.330	0.323	0.259
FI-3	Failure of government departments to recognise & understand the complexity of refinancing		1	8	4	12	0.785	1	5	0.876	0.336	0.881
FI-4	Financial bailouts that benefit PFI NHS Trusts	1	2	9	10	3	0.669	14	52	0.349	0.806	0.732
FI-5	Financial consequences resulting from changing scope of project		1	6	10	8	0.769	2	8	0.259	0.424	0.460
FI-6	Requirement for private sector to keep more than 50% of refinancing gains if making less profit	1	2	12	6	4	0.654	16	65	0.096	0.637	0.376
FI-7	Reward to private sector companies for delivering inefficiencies	2	6	9	4	4	0.592	22	87	0.424	0.450	0.420
FI-8	Treatment of PFI projects as off-balance sheet debt		4	10	8	3	0.654	16	65	0.825	0.608	0.030
FI-9	Creation of mortgage for the future generation		4	7	11	3	0.677	11	46	0.209	0.287	0.142
FI-10	Failure of lenders to withhold funding to trigger financial review of project		4	7	9	5	0.692	7	34	0.018	0.476	0.834
FI-11	Unwillingness of private sector to participate in voluntary code to share refinancing gains		6	8	5	6	0.662	15	59	0.178	0.403	0.125
FI-12	Lack of confidence in the financial models applied in the private sector	2	5	6	8	4	0.631	20	76	0.189	0.309	0.424
FI-13	Excessive return to the private sector that does not reflect level of risk	2	4	7	7	5	0.646	19	69	0.885	0.704	0.121

(Contiued)

Table 14.4 (Continued)

		Likert scales								*Kruskal Wallis test*		
		1	*2*	*3*	*4*	*5*	*RII*	*Ranking for category*	*Overall ranking*	*Experience*	*Organisation*	*Profession*
	Financial issues (*N* = 23)						*0.681 (4)*					
FI-14	Inability of private sector to operate efficiency and economically	2	9	4	5	5	0.592	22	87	0.106	0.414	0.111
FI-15	Annual charges that are far higher than traditional forms of payment		2	8	8	7	0.731	4	21	0.882	0.304	0.058
FI-16	Wide & unexplained variations in the cost of facilities management services in PFI hospitals		6	3	13	3	0.677	11	46	0.027	0.376	0.044
FI-17	Funding & budgeting mechanisms that make on balance sheet projects less attractive		3	9	9	4	0.685	10	43	1.000	0.269	0.405
F-18	Failure of public sector to approach investors & contractors to negotiate share in efficiency gains & economies of scale		2	12	5	6	0.692	7	34	0.445	0.937	0.596
F-19	Disruption in the credit markets affecting funding for PFI projects, that is pricing and available of project debt	1	2	5	12	5	0.715	6	27	0.569	0.567	0.709
F-20	Failure of private sector disclosing or notifying the public sector of refinancing	1	7	7	6	4	0.615	21	78	0.610	0.599	0.174
F-21	Need for SPV to hold surplus cash to meet the requirement of lenders	1	2	10	5	7	0.692	7	34	0.442	0.511	0.629
F-22	Private sector transfer of insurance cost (buildings & business interruption) to the public sector		5	11	3	6	0.654	16	65	0.485	0.209	0.090
F-23	Difficulty terminating PFI contracts, that is rate swaps, upfront funding		4	5	8	8	0.731	4	21	0.213	0.307	0.126

Source: Original.

Table 14.5 Joint display of qualitative and quantitative data (factors that affect the reputation of PFI projects)

																Likert scales								*Kruskal Wallis test*		
Codes	*Doc-1*	*Doc-2*	*Doc-3*	*Doc-4*	*Doc-5*	*Doc-6*	*Doc-7*	*Doc-8*	*Doc-9*	*Doc-10*	*Doc-11*	*Doc-12*	*Doc-13*	*Doc-14*	*%*	*1*	*2*	*3*	*4*	*5*	*RII* 0.681 (4)	*Ranking for category*	*Overall ranking*	*Experience*	*Organisation*	*Profession*
FI-1		X								X					14		2	6	8	9	0.762	3	9	0.793	0.139	0.390
FI-2	X			X											14		3	6	11	4	0.677	11	46	0.330	0.323	0.259
FI-3														X	7		1	8	4	12	0.785	1	5	0.876	0.336	0.881
FI-4			X	X											14	1	2	9	10	3	0.669	14	52	0.349	0.806	0.732
FI-5				X											7		1	6	10	8	0.769	2	8	0.259	0.424	0.460
FI-6														X	7	1	2	12	6	4	0.654	16	65	0.096	0.637	0.376
FI-7				X											7	2	6	9	4	4	0.592	22	87	0.424	0.450	0.420
FI-8						X		X					X		21		4	10	8	3	0.654	16	65	0.825	0.608	0.030
FI-9													X		7		4	7	11	3	0.677	11	46	0.209	0.287	0.142
FI-10				X								X			14		4	7	9	5	0.692	7	34	0.018	0.476	0.834
FI-11														X	7		6	8	5	6	0.662	15	59	0.178	0.403	0.125
FI-12				X											7	2	5	6	8	4	0.631	20	76	0.189	0.309	0.424
FI-13					X	X	X				X	X			36	2	4	7	7	5	0.646	19	69	0.885	0.704	0.121
FI-14				X											7	2	9	4	5	5	0.592	22	87	0.106	0.414	0.111
FI-15						X							X		14		2	8	8	7	0.731	4	21	0.882	0.304	0.058
FI-16					X					X					14		6	3	13	3	0.677	11	46	0.027	0.376	0.044
FI-17					X			X					X		21		3	9	9	4	0.685	10	43	1.000	0.269	0.405

Source: Original.

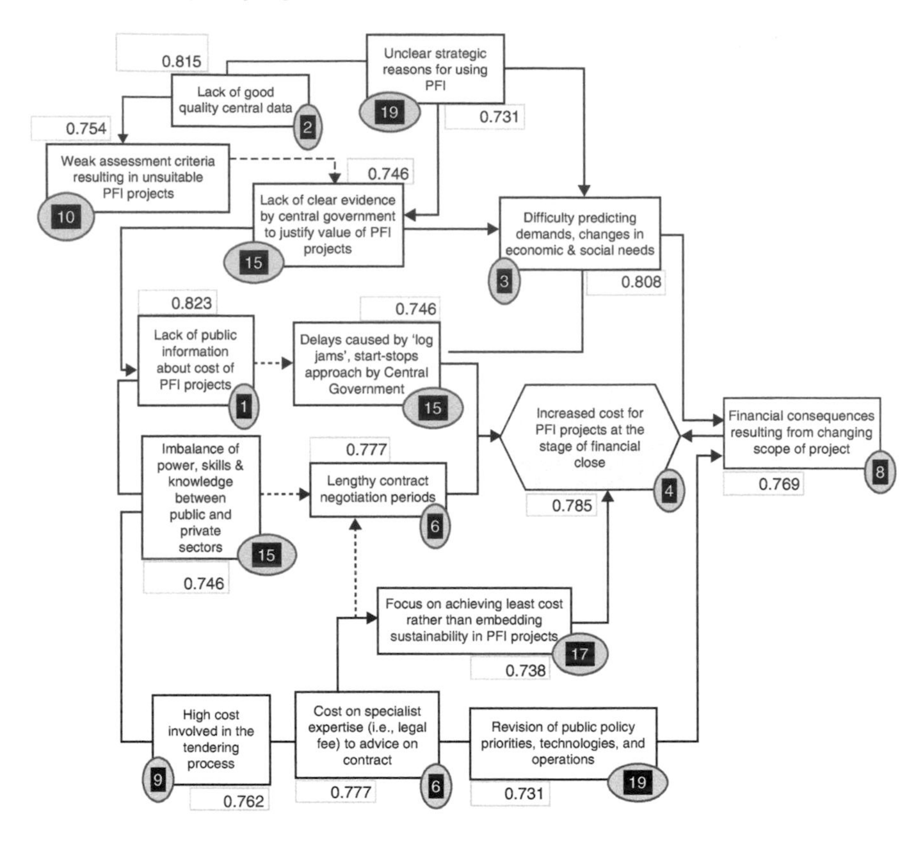

Figure 14.3 Theoretical frameworks of factors that affect the reputation of PFI projects

the list of factors under "financial category" was used in the development of the questionnaire and analysed using RII and the Kruskal Wallis est. This represents the first point of integration that is crucial for qualifying mixed methods research (Schoonenboom & Johnson, 2017).

The second point of integrating the qualitative and quantitative research occurred at the results interpretation level to answer the research questions. The results of the qualitative as well as quantitative phases of this research are presented using a joint display in Table 14.5. Joint display makes it possible for data to be visually brought together (Berman, 2017) to "draw out new insights beyond the information gained from the separate quantitative and qualitative results" (Berman, 2017, cited by Fetters *et al.*, 2013: 2143). As shown in Table 14.5, the results of the exploratory content analysis are presented side by side with those of the RII. In terms of exploratory content analysis, "excessive return to the private sector that does not reflect risk" was ranked the most important factor (36%, F13). With RII of 0.646, this factor was ranked 19 out of the 23 factors in the financial category; it was also ranked 69 out of 91 factors for all categories. Whilst the research of the content analysis was mainly exploratory in nature, it provided useful information about how some of these issues were viewed by different authors.

The refined list of factors and results of the RII were used in the development of the theoretical framework of factors that affect the reputation of PFI/P2 projects in the UK. Apart from merely showing the relation between the different factors, using RII also made it possible to show their level of significance. The theoretical framework also provided room for the qualitative as well as quantitative results to be displayed jointly. Only the most important factors (20) drawn from the categories were used here. As shown in Figure 14.3, the numbers in the circles represent the overall rank of the factors. This shows, for example, the link between "lack of public information about the true cost of PFI/PF2 projects" and "imbalance of power, skills and knowledge between the private and public sectors".

Discussion of Findings and Implications of the Research Design

Mixed methods research involves separate analysis of qualitative and quantitative data to address the research questions. In the first phase of this research, relevant research material was analysed using the Straussian GT, involving open, axial, and selective coding. The themes contained in the different nodes were analysed and refined for the development of the research instrument. In total, 91 factors were identified and grouped into seven parent nodes. The results of the qualitative content analysis were mainly exploratory based on an interrogation of literature about issues that affect the reputation of PFI projects in the UK. The results suggested that these issues had not been given the right attention. Concerning the quantitative phase of the research, data generated from the administration of a questionnaire survey were analysed using RII and the Kruskal Wallis Test. This constituted an important point of integrating the qualitative and quantitative results. According to Berman (2017: 7), "the use of both qualitative and quantitative data collection methods in a single study is not sufficient enough to categorize a study as 'mixed methods'". The true value of mixed research lies in the mixing of the two strands of data.

The chapter also showed how the results of the qualitative and quantitative phases of the research could be jointly displayed to bring more meaning to the phenomena under discussion. This is an important characteristic and strength of mixed methods research. In this case, the refined list of the factors, results of the exploratory content analysis and RII were presented side-by-side. The qualitative results provided useful information about how these issues are addressed in the literature. On the other hand, the quantitative results provided information about their level of significance. Both sets of results were then used to develop a theoretical framework of factors that affect the reputation of PFI/PF2 projects in the UK.

Conclusion

The chapter contains a discussion of different steps involved in exploratory sequential mixed methods research used to investigate factors that affect the reputation of PFI projects in the UK. In the chapter, the process involved in selecting documents for conducting GT and applying open, axial and selected coding was shown. The use of open coding enabled the identification of a list of factors that affect the reputation of PFI/PF2 projects in the UK. Through iteration in axial coding, the factors were categorised and the coded material was merged. For example, the parent nodes of finance and re-finance were merged to provide clarity. In total, the process resulted in the identification of seven categories containing 91 factors. The lists of factors in the different categories were used to develop a questionnaire survey that was despatched to selected individuals in the construction industry.

One of the strengths of the mixed research methodology is the integration and joint display of the results. The first level of integration of the qualitative and quantitative research occurred when the refined list of factors that was developed using GT analysis, was transformed into a survey questionnaire. The data were analysed using RII and jointly displayed with the results of the exploratory content analysis. Tables 14.4 and 14.5 show the most important factors identified using exploratory content analysis and RII. Both results were then used to inform the selective coding process for the development of the theoretical framework of factors that affect the reputation of PFI/PF2 projects in the UK.

References

Abowitz, D.A., Toole, T.M. 2010. Mixed method research: Fundamental issues of design, validity, and reliability in construction research. *Journal of Construction Engineering and Management*, 136: 108–116.

Berman, E.A. 2017. An exploratory sequential mixed methods approach to understanding researchers' data management practices at UVM: Integrated findings to develop research data services. *Journal of eScience Librarianship*, 6(1): e1104. https://doi.org/10.7191/jeslib.2017.1104

Brown, I. & Roode, D. 2004. Using grounded theory in the analysis of literature: The case of strategic information systems planning. *ACIS 2004 Proceedings*, 113. http://aisel.aisnet.org/acis2004/113

Bryman, A. 1984. The debate about quantitative and qualitative research: A question of method or epistemology? *The British Journal of Sociology*, 35: 75–92.

Buber, R., Gadner, J. & Richards, L 2004. *Applying Qualitative Methods to Marketing Management Research*. Palgrave School, Houndmills, Basingstoke, Hampshire; New York.

Bulawa, P. 2014. Adapting grounded theory in qualitative research: Reflections from personal experience. *International Research in Education*, 2: 145–168.

Carrillo, P., Robinson, H., Foale, P., Anumba, C. & Bouchlaghem, D. 2008. Participation, barriers, and opportunities in PFI: The United Kingdom experience. *Journal of Management in Engineering*, 24: 138–145.

Creswell, J.W. & Clark, V.L.P. 2007. *Designing and Conducting Mixed Methods Research*. SAGE Publications, Thousand Oaks, CA.

Creswell, J.W. & Plano Clark, V.L. 2018. *Designing and Conducting Mixed Methods Research*. SAGE, Thousand Oaks, CA.

De Loo, I. & Lowe, A. 2011. Mixed methods research: Don't "just do it." *Qualitative Research in Accounting and Management*, 8: 2238.

Doyle, L., Brady, A.-M. & Byrne, G. 2009. An overview of mixed methods research. *Journal of Research in Nursing*, 14: 175–185.

Dunne, C. 2011. The place of the literature review in grounded theory research. *International Journal of Social Research Methodology*, 14: 111–124.

Edmonds, W.A. & Kennedy, T.D. 2017. *An Applied Guide to Research Designs: Quantitative, Qualitative, and Mixed Methods*. SAGE Publications, Inc, Thousand Oaks, CA.

Fetters, M.D., Curry, L.A. & Creswell, J.W. 2013. Achieving integration in mixed methods designs – Principles and practices. *Health Services Research,* 48: 2134–2156.

Goddard, W. & Melville, S. 2001. *Research Methodology: An Introduction*, 2nd edition, Juta and Company Ltd, Lansdowne.

Graff, J. C. 2017. An integrative approach to research, administration and practice. In: L.A. Roussel & H.R. Roussel (eds). *Mixed Method Research,* pp. 47–65. Jones & Bartlett Learning, Burlington, MA.

Hair, J.F., Anderson, R.E., Tatham, R.L. & Black, W.C. 1998. *Multivariate Data Analysis.* 5th edition. Prentice-Hall, Englewood Cliffs, NJ.

Hall, H.R. & Roussel, L.A. (eds). 2017. *Evidence-Based Practice: An Integrative Approach to Research, Administration, and Practice.* 2nd edition. Jones & Bartlett Learning, Burlington, MA.

Johnson, B. & Onwuegbuzie, A. 2004. Mixed methods research: A research paradigm whose time has come. *Educational Researcher*, 33(7): 14–26.

Malina, M.A., Nørreklit, H.S.O. & Selto, F.H. 2011. Lessons learned: Advantages and disadvantages of mixed method research. *Qualitative Research in Accounting & Management*, 8: 59–71.

Morse, J.M., Niehaus, L. 2009. *Mixed Method Design: Principles and Procedures, Developing Qualitative Inquiry*. Left Coast Press, Walnut Creek, CA.

Sale, J.E.M., Lohfeld, L.H. & Brazil, K. 2002. Revisiting the quantitative-qualitative debate: Implications for mixed-methods research. *Quality & Quantity,* 36: 43–53.

Schoonenboom, J. & Johnson, R.B. 2017a. How to construct a mixed methods research design. *Kolner Z Soz Sozpsychol,* 69: 107–131.

Schoonenboom, J. & Johnson, R.B., 2017b. How to construct a mixed methods research design. *Kolner Z Soz Sozpsychol,* 69: 107–131.

Shiyanbola, O.O., Rao, D., Bolt, D., Brown, C., Zhang, M. & Ward, E. 2021. Using an exploratory sequential mixed methods design to adapt an Illness Perception Questionnaire for African Americans with diabetes: The mixed data integration process. *Health and Psychological Behaviour Medicine,* 9: 796–817.

Somiah, M.K., Osei-Poku, G. & Aidoo, I. 2015. Relative importance analysis of factors influencing unauthorized siting of residential buildings in the Sekondi-Takoradi Metropolis of Ghana. *JBCPR*, 3: 117–126.

Strauss, A. & Corbin, J.M. 1990. *Basics of Qualitative Research: Grounded Theory Procedures and Techniques.* Sage Publications, Inc, Thousand Oaks, CA.

Tashakkori, A. & Creswell, J.W. 2007. Editorial: The new era of mixed methods. *Journal of Mixed Methods Research,* 1: 3–7.

Tashakkori, A. & Creswell, J.W. 2008. Editorial: Mixed methodology across disciplines. *Journal of Mixed Methods Research,* 2(1): 3–6.

Teddlie, C. & Tashakkori, A. 2009. *Foundations of Mixed Methods Research: Integrating Quantitative and Qualitative Approaches in the Social and Behavioral Sciences.* SAGE, Los Angeles.

Teddlie, C. & Tashakkori, A. 2008. *Foundations of Mixed Methods Research: Integrating Quantitative and Qualitative Approaches in the Social and Behavioral Sciences*. 1st edition. SAGE Publications, Inc, Los Angeles.

Whitefield, D. 2017. PFI/PPP buyouts, bailouts, terminations and major problem contracts in UK. *International Journal of Doctoral Studies*, 16: 553–568.

Wolfswinkel, J., Furtmueller, E. & Wilderom, C. 2013. Using grounded theory as a method for rigorously reviewing literature. *European Journal of Information Systems,* 4: 22.

Yu, M. & Smith, S.M. 2021. Grounded theory: A guide for a new generation of researchers. *International Journal of Doctoral Studies*, 16: 553–568.

15 An Exploration of the Implications of Sustainable Construction Practice

Mixed Methods Research Approach

Cheng Siew Goh and Shamy Yi Min Chin

Summary

Sustainable construction involves multi-disciplinary work from diverse stakeholders. The complex and complicated relationships between different actors require a more thorough examination in assessing sustainability performance. The complexity necessitates adopting a holistic perspective to appreciate the multi-dimensional realities of sustainability using mixed methods. Mixed methods offer different methodological primacy and offset weaknesses associated with stand-alone quantitative or qualitative methods. In this chapter, a study is presented to demonstrate how quantitative data from questionnaire surveys and qualitative data from interviews can be integrated to give a more holistic view of the implementation of the three pillars of sustainable construction. The aim of the study was to explore the stakeholder perceptions of adopting the three pillars of sustainability in construction and examine their implications. A concurrent mixed methods design was adopted for the study. Both qualitative and quantitative results suggested the disproportionate implementation of the environmental, social and economic pillars in the pursuit of sustainability. The use of mixed methods provides a richer understanding of complex sustainability issues. There are considerable benefits of using mixed methods to appreciate the "truth" by taking a synergistic view from multiple perspectives.

Introduction

The complexity level of research within the built environment could be further heightened when considering sustainable construction. Sustainable construction has been advocated increasingly in response to climate change and environmental degradation (Kibert, 2007). Transforming the built environment according to sustainable practice has been given great emphasis in view of the undeniable significance of buildings in contributing to environmental changes, social development and economic growth (Goh, 2017; Shi *et al.*, 2012). Since then, there has been a shift from the traditional mindset to a more balanced, sustainability paradigm and this drives the concept of sustainable construction in the global and local construction agenda (Chang *et al.*, 2018; Goh, 2017).

DOI: 10.1201/9781003204046-15

Sustainability in the built environment involves extensive physical boundaries and consideration of the impacts of the entire life cycle, ranging from material extraction to manufacturing, construction, operation and disposal on the health of the human and natural system (Kibert, 2007). Apart from environmental footprints, sustainability is linked to social changes and economic implications. A key challenge to comprehending sustainability has been the problem of the incommensurability of different dimensions or strategies of sustainable construction practice (Kibert, 2007). Edum-Fotwe and Price (2009) concurred and highlighted the complex interaction of a multiplicity of factors relating to sustainability and their associated challenges in defining the multiplicity. As a result, it is not uncommon to find sustainability studies focusing predominantly on the environmental dimension of sustainability, with the social dimension being addressed less (Edum-Fotwe & Price, 2009).

Meanwhile, Araújo *et al.* (2020) conducted a systematic review of sustainability studies in construction and found that mostly quantitative methodologies were used to identify aspects of sustainability, to create/adopt indicators for specific contexts, as well as to classify and assess construction projects. There is a lack of sufficiency in current studies defining social-related factors which embrace a greater proportion of subjective factors for consideration (Edum-Fotwe & Price, 2009). Considering the varying nature of the aspects and indicators of sustainability, the evaluation of sustainability using a stand-alone research methodology would fall short of converting different considerations, units and measurements into one consolidated context.

Multi-disciplinary and inter-disciplinary topics embedded in sustainable development require prospecting (Araújo *et al.*, 2020). To permit a more rigorous analysis and to capture dynamic changes of various constructs, ranging from environmental to social, economic and technological aspects, it is crucial to adopt a more systematic approach to appreciate the realities of sustainable construction. Mixed methods could help bridge the gaps in examining the central issues of environment, economy and society in the sustainable built environment.

In this chapter, a study is presented to demonstrate how quantitative data from questionnaire surveys and qualitative data from interviews can be integrated to give a more comprehensive view of the implementation of the three pillars of sustainable construction. The aim of the study was to explore the stakeholders' perceptions of the implications of adopting the three pillars of sustainability in construction. The objective of the study was to identify challenges in implementing the three pillars of sustainability in construction practice. The use of mixed methods gives multiple perspectives towards understanding sustainable construction.

Sustainable Construction

Sustainability is defined as the ability to meet current needs without compromising the needs of future generations (WCED, 1987). It entails the three pillars, that is the environmental, social and economic aspects of sustainability. Since the vision of sustainability is to embrace all the three pillars in development, a wide range of factors and aspects shall be taken into consideration. As propagated in the United Nation's 2030 Agenda, the 17 Sustainable Development Goals call for capacity-building in the spheres of water, energy, climate, urbanisation, transport, science and technology. As a result, the implementation of sustainable construction involves multi-disciplinary works and entails various perspectives, ranging from environmental, social and economic to technological factors.

Although sustainability rests on the three pillars, literature shows that the environmental, social and economic aspects of sustainability are not adopted at a similar level in the construction industry. Existing studies are found to focus predominantly on environmental impacts, particularly energy consumption, greenhouse gas emissions, water efficiency and renewable energy (Chwieduk, 2003; Fowler *et al.*, 2010; Goh *et al.*, 2020; Newsham *et al.*, 2009; Zou & Zhao, 2014). Studies on social and economic sustainability are found to be relatively lean (Zou & Zhao, 2014). The tendency of environmental studies could be linked to quantifiable factors and measurements involved in the environmental aspect, hence potentially increasing the adoption of quantitative methodology for environmental sustainability studies.

There have also been extensive research works in which the implications of adopting sustainability in construction practice have been investigated. However, the impacts on environment, society and economy have been examined in a collective manner in only a few of these studies. The environmental implications of sustainable construction are greatly acknowledged because of the disproportionality of research in sustainability on the environmental pillar. People have an increasing awareness of the importance of sustainable construction in contributing to the environment such as reductions of carbon emission, pollution, resource consumption and waste generation. With proper management and monitoring strategies, sustainable construction will be able to help to mitigate climate change and restore biodiversity and ecosystems.

On the other hand, the social and economic implications of sustainability are not as widely discussed in the literature as the environmental implications. In previous studies, attempts were made to explore the economic implications by evaluating the costs and benefits of sustainable construction projects. For instance, Gan *et al.* (2015) reported that sustainable buildings will incur higher initial costs compared with non-sustainable buildings because of the incorporation of green technology, integrated design, sustainable building certifications and special arrangement for administering and operating sustainable buildings. In the same study, the economic benefits of sustainable construction were also highlighted, as the projects are often associated with high investment and long payback periods. As for social implications, sustainable practice helps the relevant organisations and projects to gain a positive reputation, with an improvement in community relations. Improved user health, comfort and satisfaction are also recognised as social implications resulting from sustainable construction projects.

There is apparently a gap in addressing the totality of the implications when assessing the sustainable built environment. As stated by Goh *et al.* (2020), there has been a lack of in-depth investigation of sustainable construction from the perspective of environmental, economic and social sustainability as a whole. In some studies, attempts have been made to examine the environmental, social and economic aspects of sustainable construction practice (Beheiry *et al.*, 2006) but their analyses of these aspects were often conducted individually. Studies of sustainable development in construction require a more integrated research approach to lead to a more nuanced understanding of various processes and agents in the diffusion and implementation of the practice of sustainability in the built environment.

A synergy between the three pillars is essential to achieve the sustainability goals. Sustainable construction involves multi-disciplinary work with diverse stakeholders. The complex and complicated relationships between different actors often require a more thorough examination in assessing performance in sustainability. Therefore, the gap is filled in this research study by investigating the environmental, social and economic implications of applying the three pillars in the built environment in a collective manner.

Methodology

Day and Gunderson (2018) proposed that built environment research problems require a combination of research methods spanning across different disciplines. As such, mixed methods that draw upon both quantitative and qualitative research are appropriate. Similarly, research in sustainable construction calls for a more consolidated research approach that transcends different methods to explain social phenomena. A blend of research methods can prompt researchers to link judgement and analysis to give a complete picture of sustainable construction practice.

Using a single methodological standpoint could lead to some unknown or restrictive aspects of the results (Robson, 1993). There is an over-reliance on quantitative research in past studies of the built environment (Dainty, 2008). Although qualitative research has become slightly influential in recent years, its adoption was not been well received in the past as people often viewed it as lacking rigour and objectivity (Boddy, 2016). Therefore, mixed methods research is used to fill the gaps associated with sustainability in the built environment.

A quantitative perspective is helpful to provide a general understanding of the implementation of sustainable construction to inform stakeholders where they stand in the transition to sustainability. Quantitative methods can provide a good analysis of the trend of sustainable construction processes with predictive modelling, hence reducing uncertainties in sustainable construction (Araújo *et al.*, 2020). However, quantitative methods have limited ability to account for complex interactions between variables (Korkmaz *et al.*, 2011). Meanwhile, qualitative methods allow for in-depth analysis of complex phenomena and behaviours in a natural setting by obtaining first-hand data on issues of interest and the project situations (Korkmaz *et al.*, 2011). Qualitative perspectives are critical to explore the underlying reasons behind the observed phenomena to identify what needs to be done in the next steps. Qualitative approaches provide a naturalistic setting to help understand the state of implementation of the three pillars and the underlying factors of such implementation. The objective of the study was to explore the stakeholders' perceptions of adopting the three pillars of sustainability in construction and to examine the implications of such adoption. In this study, concurrent mixed methods design was used to support the achievement of the research objectives at different levels.

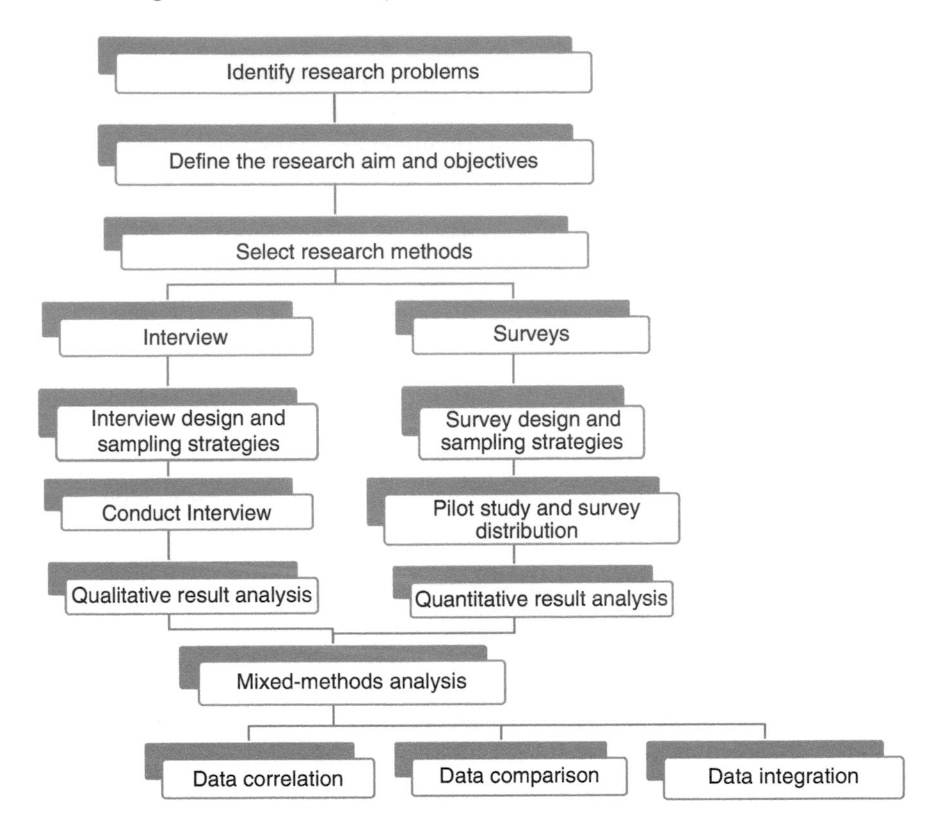

Figure 15.1 Flowchart of the research process

Figure 15.1 illustrates the procedures undertaken in the study. The selection of research methodology was guided by the research aim and objectives. The purpose of the study was to investigate the magnitude of sustainability implementation and describe stakeholders' perceptions of the contemporary practice of sustainable construction. As a result, the study required both qualitative and quantitative forms of data. Interviews were used to collect qualitative data, while a survey was adopted to collect quantitative evidence. Collection of qualitative and quantitative data was carried out at the same time, with no enquiry of either approach being used to inform the development of the research instrument for the other approach.

Time orientation and the relationship between the collection of quantitative and qualitative data are the core elements for determining the types of mixed methods design models adopted in a study. A concurrent design model is efficient because the collection of qualitative and quantitative data is collected during the same period. The collected data were first analysed separately using the conventional qualitative or quantitative analysis methods. Integration of qualitative and quantitative data was then done at the data analysis and interpretation levels. Integration of qualitative and quantitative data was conducted in two ways: (1) by articulating the findings in a distinct section within which results of qualitative and quantitative data were discussed separately; (2) by presenting the data in a graphic form of either a table or figure that displays quantitative and qualitative results (a joint display). The adoption of the concurrent design model reduced the time spent on data collection.

Quantitative Method: Design and Sampling Strategies

The questionnaire included: (1) demographic questions, (2) perceptions relating to the implementation of the three pillars of sustainability in construction and (3) the perceptions of stakeholders concerning the implications and challenges of adopting these pillars. The questionnaire contained different types of questions such as multiple choice, Likert Scale ranking and short answers. A five-point Likert Scale was used to measure the state of implementation and to indicate the extent of the respondents' agreement with a provided statement.

The survey was conducted in two forms: electronic questionnaires and self-administered questionnaires. Electronic questionnaires were sent to respondents by email, while self-administered questionnaires were distributed to respondents associated with professional organisations. A total of 129 questionnaires were distributed and 42 valid responses were received, giving a response rate of 32.56%. Descriptive statistics were used to examine the distributional properties of the studied variables. Frequency distribution was preferred because the quantitative results facilitated data integration by complementing the thematic analysis findings.

Qualitative Method: Design and Sampling Strategies

Semi-structured interviews with experts in the study area were conducted. The interview questions were designed to elicit the perceptions of interviewees concerning the adoption of three pillars of sustainability in construction practice and the reasons informing such perceptions.

Purposive sampling was used to select interviewees based on their ability to provide rich information on the studied phenomenon – the adoption of the three pillars in sustainable construction. Although a qualitative research sample is usually small, the depth and robustness of qualitative data can help to assess the state of implementing sustainable construction in a comprehensive manner. Two interview sessions were conducted. Similarity found in the interview excerpts suggested that a saturation point had been attained.

Results and Discussion

The analyses of the results of the quantitative and qualitative data are presented separately, followed by an integrated presentation in the following sections.

Quantitative Strands: Results and Analysis

The quantitative results showed that there is a generally high level (~81%) of implementing all three pillars of sustainability: environmental, social and economic sustainability in the construction industry. As shown in Table 15.1, the environmental pillar had the highest mean score (3.33), followed by the social pillar (2.90) and the economic pillar (2.83). The results concur with previous studies (Edum-Fotwe & Price, 2009; Goh *et al.*, 2020; Zou & Zhao, 2016) that show the environmental pillar is always given priority while the other two pillars are overlooked.

Respondents were also asked to indicate the initiatives they implemented for the three pillars of sustainability in their practice. Over 71% and 64% of respondents implemented initiatives that encourage low energy consumption and the use of renewable/recyclable resources respectively. The mid-range implementation rating of 48% was indicated for only one initiative for the economic pillar and two initiatives for the social pillar. Table 15.2 shows the initiatives with regard to the three pillars of sustainability that were implemented by respondents.

Figure 15.2 shows the frequency distribution of implications of implementing the three pillars of sustainability. Most of the respondents (over 80%) found that the implementation of the three pillars can lead to a healthier built environment.

Table 15.1 Level of adoption of the three pillars of sustainable construction practice

	Never	*Rarely*	*Sometime*	*Often*	*Very often*	*Mean score*
Environment	0	0	6	18	10	3.333
Economy	0	2	15	15	2	2.833
Society	0	4	8	20	2	2.905

Source: Original.

Table 15.2 Initiatives taken for the implementation of the three pillars of sustainability

The pillar of sustainability	*Sustainable construction initiatives*	*Implemented by respondents (%)*
Environmental pillar	Initiatives encouraging low energy consumption	71
Environmental pillar	Initiatives using renewable or recyclable resources	64
Economic pillar	Initiatives reducing operational and maintenance cost (or emphasising life cycle cost)	48
Social pillar	Initiatives allowing community participation to promote open spaces and increase user interactions	48
Social pillar	Initiatives considering users' needs and comfort in design and construction	48
Environmental pillar	Initiatives integrating environmental thinking into the project supply chain	45
Environmental pillar	Initiatives in relation to waste management	38
Economic pillar	Initiatives to ensure sustainable profits for construction businesses	26
Social pillar	Initiatives adopting collaborative or partnering approaches	26
Social pillar	Initiatives encouraging local purchase and local labour policies	24

Source: Original.

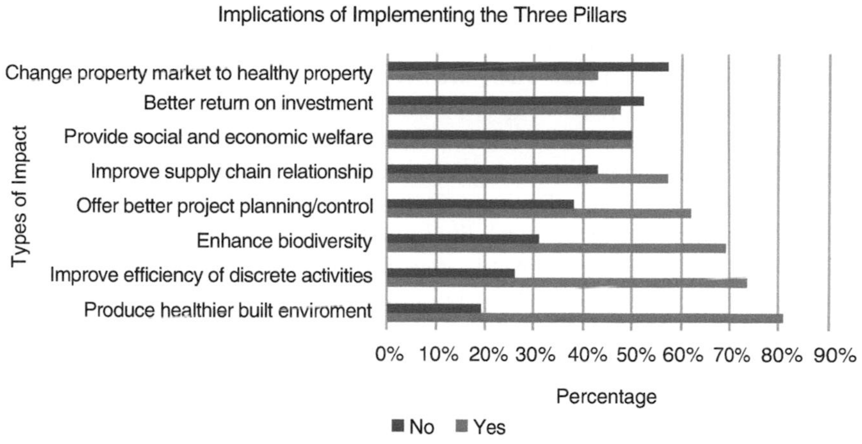

Figure 15.2 Implications of implementing the three pillars of sustainability in construction practice

Table 15.3 Challenges of implementing the three pillars of sustainability

Challenges	*Rating*					*Mean score*
	1	*2*	*3*	*4*	*5*	
Stakeholder awareness	0	1	8	15	18	4.190476
Lack of government commitment	1	1	11	18	11	3.880952
Significant cost incurred	1	2	10	21	8	3.785714
Inadequate professionals	0	4	13	17	8	3.690476
Market demand	0	5	18	14	5	3.452381
Insufficient resources	0	8	13	18	3	3.380952

Source: Original.

There was also consensus that the three pillars of sustainability can increase the efficiency of construction activities (74%) and enhance the biodiversity of the built environment (69%). There was a low level of agreement from respondents that the three pillars contribute to a good return on investment in the property market, with 43% and 48% agreement from respondents, respectively.

In Table 15.3, the awareness of the three pillars of sustainability (mean score of 4.19) was recognised as the main challenge to the adoption of the three pillars in construction practice, followed by a lack of government commitment (mean score of 3.88) and significant cost incurred (mean score of 3.78), inadequate sustainability professionals (mean score of 3.69), a lack of market demands (mean score of 3.45) and insufficient resources (mean score of 3.38).

Qualitative Strand: Findings and Analysis

Qualitative data was analysed using thematic analysis. The interview findings revealed that both interviewees had adopted the three pillars of sustainability in their practice. Nevertheless, both interviewees were unsatisfied with the current practice of sustainability in construction. Interviewee 1 felt that practising sustainability in construction is a challenging task. There has been a lack of professionals with relevant experience in dealing with sustainability in construction. Nevertheless, Interviewee 1 observed that the practice of the three pillars has created positive impacts for their organisations, with increasing chances to win construction bids.

From the perspective of Interviewee 2, there had been an imbalance in the adoption of the three pillars of sustainability in practice. While the environmental and social pillars were widely adopted in construction projects, Interviewee 2 found that the economic pillar had a relatively lower level of application when compared with the others. There appears to be a lack of consideration of the economic pillar in sustainable construction projects and some of the property projects involved were seen to have incurred a higher price. Interviewee 2 concurred with Interviewee 1 regarding the usefulness of adopting the three pillars in helping to create a positive impact.

Interviewee 2 perceived a lack of sustainability professionals as being the main challenge of practising the three pillars of sustainability. Despite their awareness, construction stakeholders did not possess high-level skills and specialisation in sustainability. Interviewee 1 found the inadequacy of resources to be the main challenge.

Integration of Qualitative and Quantitative Data

Integration of qualitative and quantitative data was conducted in two ways: (1) integrating in the discussion and (2) integrating using joint displays. A joint display provides a method and cognitive framework to bring the mixed methods data together using visual means to draw out new insights from the quantitative and qualitative results (Guetterman *et al*., 2015). Analysis integrated into the discussion and joint displays is presented below,

Presentation of Findings

As shown in Figure 15.3, both results suggested that the development of the three pillars in sustainable construction practice is disproportionate. The quantitative data showed that the environmental pillar had the highest weighting, followed by the social and economic pillars. The result was confirmed by the qualitative data

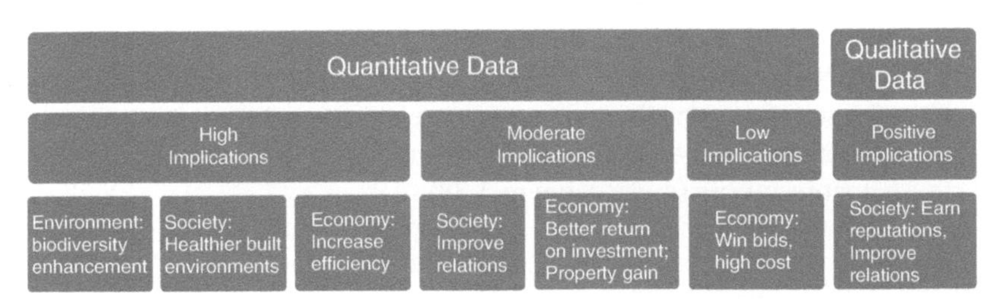

Figure 15.3 Complementary results from quantitative and qualitative data on the implications of practising the three pillars

that showed that the economic pillar was neglected by practitioners, compared with the other two pillars, in their pursuits of sustainability.

With regard to the challenges of adopting the three pillars, the qualitative and quantitative results also showed a similar pattern in the outcome of the analysis. The main challenges identified included a lack of knowledge and experience in sustainability, a lack of the government's commitment, a lack of sustainability professionals, incurring additional costs and inadequate resources in the markets. The quantitative results corroborated the main challenges identified in the qualitative data.

However, there were some slight differences between the qualitative and quantitative results in examining the implications of practising the three pillars. The interviewees perceived the values and impacts of adopting sustainability differently, considering their professional backgrounds and conflict of interest. On one hand, the qualitative findings highlighted that the application of the three pillars could improve the company's reputation and, in turn, increase the project and organisational values. On the other hand, the quantitative results indicated that positive economic implications, such as the return on investment, and increased property values were not highly rated by stakeholders. Figure 15.3 shows the complementary results for the implications of adopting the three pillars by mapping the quantitative results with the qualitative results. It is important to note that the economic values derived from embedding the three pillars of sustainability can be overturned if the sustainability practice was not adopted effectively in the very beginning of the projects, as revealed in the qualitative findings. Stakeholders failed to consider the economic pillar in the current sustainability practice, leading to additional costs. This was probably the main reason why the return on investment was given a lower rating in the quantitative results.

Integration Using Joint Displays

Side-by-side displays were adopted for this study as the design model to present quantitative and qualitative data together using the identified themes. The themes accompanied by the quantitative and qualitative data are presented in Table 15.4.

Having the qualitative and quantitative data organised based on thematic relevance, convergence of the data can be observed. Integration of data helps to show coherence between qualitative and quantitative findings. Both qualitative and quantitative results confirmed that the environmental pillar has widespread implementation in sustainable construction practice, followed by the social pillar and the economic pillar. The results also suggested that lack of awareness and knowledge, cost increment, lack of sustainability experts, lack of government commitment and inadequate resources are the main challenges of implementing the three pillars. The confirmatory results of the two data sources enhance the reliability and credibility of the findings.

Complementary results of the qualitative and quantitative data for the implications of the three pillars are illustrated in Figure 15.3.

As presented in Table 15.4, the quantitative results suggested a high level of implementing the three pillars, while the interview results indicated an

Table 15.4 Side-by-side display of qualitative and quantitative findings

Themes/domains	*Quantitative data*	*Qualitative data*
The implementation of the three pillars of sustainability in construction.	A high level of implementing the three pillars of sustainability (as shown in Table 15.1).	The three pillars are adopted in the current practice, but the interviewees perceived the implementation level as being unsatisfactory.
The implementation level of the three pillars of sustainability.	The environmental pillar was scored the highest, followed by the social and the economic pillar (as shown in Table 15.2). High level: 2 environmental initiatives. Moderate level: 2 social initiatives, 1 economic initiative, 1 environmental initiative. Low level: 2 social initiatives, 1 economic initiative, 1 environmental initiative.	The environmental pillar and the social pillar are widely adopted but there is minimal implementation of the economic pillar.
Challenges of implementing the three pillars	Stakeholder awareness (MS 4.19), lack of government commitment, significant cost incurred (MS 3.78), inadequate professionals (MS 3.69), market demand (MS 3.45), insufficient resources (MS 3.38) (as shown in Table 15.3)	Lack of knowledge and experience in sustainability, lack of government commitment, lack of sustainability experts, inadequate resources, additional cost incurred.
Implications of implementing the three pillars.	Produce a healthier environment (81%), increase efficiency of construction activities (74%), improve biodiversity of the built environment (69%), offer better project planning/control (62%), improve supply chain relationship (57%), social and economic welfare (50%), better return on investment (48%), a healthy property market (43%) (as shown in Figure 15.2).	Positive impacts on the organisations involved, good reputation, higher chances of winning bids in the future, improved working relationship among the project team members.

Source: Original.

unsatisfactory level of implementation. The contrary results can be explained by correlating the outcomes with other responses. As revealed by the interviewees, the rate of adoption of the three pillars of sustainability is not equal, as the economic pillar was overlooked. The high level of implementation suggested by the quantitative results could refer to the widespread implementation of environmental sustainability initiatives. However, the meeting of sustainable development goals might be compromised if only one or two pillars of sustainability are practised. Lack of an equal effort in developing all the three pillars could be attributed to unsatisfactory implementation of sustainability, as perceived by the interviewees.

Implications for Mixed Methods in Sustainable Construction Research

Research in sustainable construction involves technological and human factors, and it is critical to advocate greater use of mixed methods to integrate different realms of theory and practice to diversify the knowledge generated by sustainable construction studies. The gaps in sustainable construction are found to be associated with the multiplicity and multi-dimensions of sustainability because of incommensurable factors. Despite the growing volume of literature in mixed methods research, there is limited empirical research in which the concurrent design framework is being operationalised in sustainable construction studies.

The study presented shows that mixed methods research can offer a more enriched account to uncover the extent to which the three sustainability pillars are adopted in sustainable construction practice. By augmenting qualitative data with quantitative data, the findings show that the environmental pillar received the highest attention, with several environmental initiatives being widely adopted in the current sustainable practice. By cross-examining the qualitative and quantitative data, it was affirmed in this study that the environmental pillar still has broader application in sustainability initiatives, compared with the other two pillars. The integration of the qualitative and quantitative research findings has helped to increase the consistency of the findings of the study.

An exemplar of the integration of qualitative and quantitative data analysis is demonstrated in the chapter. The two ways of integration, particularly using joint displays, are seen as a useful way of advancing integration and enhancing the interpretation of mixed methods data. Mapping the quantitative data to corresponding concepts of qualitative data confirms that the qualitative findings corroborate the analysis results. For instance, there is a synergy between qualitative and quantitative data in assessing the implications of sustainability. Positive implications identified in the qualitative findings are affirmed by quantitative data. Quantitative data give an objective measurement to explain the implications of sustainability with identified threshold levels, thus expanding the understanding of the implications of adopting sustainability in construction practice. This study can serve as a point of departure to undertake mixed methods in diversifying knowledge generation and facilitating the integration of various factors relating to sustainability.

Conclusion

Studies of sustainable construction are multi-faceted and trans-disciplinary in nature. Interactions between various stakeholders and actors in the pursuit of sustainability would also give rise to varying degrees of implications in aspects of environment, society and economy. The complex and complicated relationships between different actors in sustainable construction require a more thorough examination that might not be achieved using a single method.

Mixed methods research designs offer different methodological primacy and offset weaknesses associated with stand-alone, quantitative or qualitative methods.

In this case, the implications of adopting the three pillars can be assessed through an integration of the quantitative and qualitative findings. In this chapter, it is demonstrated how quantitative data from questionnaire surveys and qualitative data from interviews can be integrated using discussion and joint displays to give a more holistic view of the implementation of the three pillars of sustainable construction.

The convergence of qualitative and quantitative data affirmed the context of implementing the three pillars in the present study. With different perspectives accruing from qualitative and quantitative data, mixed methods yielded a full picture of implementing the three pillars, thus making the findings more reliable and generalisable. The interview findings were found to complement the questionnaire results. It is evident that utilising mixed methods offers considerable value to appreciating the "truth" by giving a synergistic view resulting from multiple perspectives. It is suggested that in future studies integration is considered at different levels such as integration at the study design level and integration at the methods level to appraise the mixed methods results.

References

Araújo, A.G., Carneiro, A.M.P. & Palha, R.P. 2020. Sustainable construction management: A systematic review of the literature with meta-analysis. *Journal of Cleaner Production*, 256: 120350.

Alavi, H. & Hąbek, P. 2016. Addressing research design problem in mixed methods research. *Management Systems in Production Engineering*, 21(1): 62–66.

Azhar, S. Ahmad, I. & Sein, M.K. 2008. Action research as a proactive research method for construction engineering and management. *Journal of Construction Engineering and Management*, 136(1): 87–98.

Babbie, E. 2013. *The Basics of Social Research*. Cengage Learning, Temecula, CA.

Blee, K.M. & Taylor, V. 2002. Semi-structured interviewing in social movement research. *Methods of Social Movement Research*, 16: 92–117.

Boddy, C.R. 2016. Sample size for qualitative research. *Qualitative Market Research: An International Journal,* 19(4): 426–432.

Bryman, A. 2008. *Social Research Methods*. 3rd edition. Oxford University Press, Oxford.

Chang, R.D., Zuo, J., Zhao, Z., Soebarto, V., Lu, Y., Zillante, G. & Gan, X.L. 2018. Sustainability attitude and performance of construction enterprises: A China study. *Journal of Cleaner Production*, 172: 1440–1451.

Collins, K.M., Onwuegbuzie, A.J. & Jiao, Q.G. 2006. Prevalence of mixed-methods sampling designs in social science research. *Evaluation & Research in Education*, 19(2): 83–101.

Dainty, A. 2008. Methodological pluralism in construction management. In: A. Knight & L. Ruddock (eds). *Advanced Research Methods in the Built Environment*, pp. 1–12. Blackwell Publishing Ltd, Oxford.

Day, J.K. & Gunderson, D. 2018. Mixed methods in built environment research. *54th ASC Annual International Conference Proceedings*, Texas. http//www.ascpro.ascweb.org.

Edum-Fotwe, F.T. & Price, A.D. 2009. A social ontology for appraising sustainability of construction projects and developments. *International Journal of Project Management*, 27(4): 313–322.

Fetters, M.D., Curry, L.A. & Creswell, J.W. 2013. Achieving integration in mixed methods designs – Principles and practices. *Health Services Research*, 48(6 pt2): 2134–2156.

Goh, C.S. 2017. Towards an integrated approach for assessing triple bottom line in the built environment. Paper presented at SB-LAB 2017 – International Conference on Advances on Sustainable Cities and Buildings Development, Portugal.

Goh, C.S., Chong, H.Y., Jack, L. & Faris, A.F.M. 2020. Revisiting triple bottom line within the context of sustainable construction: A systematic review. *Journal of Cleaner Production*, 252: 119884.

Guetterman, T.C., Fetters, M.D. & Creswell, J.W. 2015. Integrating quantitative and qualitative results in health science mixed methods research through joint displays. *The Annals of Family Medicine*, 13(6): 554–561.

Hogain, D.O.C. 2018. Mixed methods research: A methodology in social sciences research for the construction manager. *Proceedings of ARCOM Doctoral Workshop Research Methodology*, https://doi.org/10.21427/rvmq-j038.

Kibert, C.J. 2007. The next generation of sustainable construction. *Building Research & Information*, 35(6): 595–601.

Korkmaz, S., Riley, D. & Horman, M. 2011. Assessing project delivery for sustainable, high-performance buildings through mixed methods. *Architectural Engineering and Design Management*, 7(4): 266–274.

Lincoln, Y.S., Lynham, S.A. & Guba, E.G. 2011. Paradigmatic controversies, contradictions, and emerging confluences, revisited. *The Sage Handbook of Qualitative Research*, 4(2): 97–128.

Newsham, G.R., Mancini, S. & Birt, B.J. 2009. Do LEED-certified buildings save energy? Yes, but…. *Energy and Buildings*, 41(8): 897–905.

Schweber, L. 2015. Putting theory to work: The use of theory in construction research. *Construction Management and Economics*, 33(10): 840–860.

Shi, Q., Zuo, J. & Zillante, G. 2012. Exploring the management of sustainable construction at the programme level: A Chinese case study. *Construction Management and Economics*, 30(6): 425–440.

WCED, S.W.S. 1987. World commission on environment and development. *Our Common Future*, 17(1): 1–91.

Zou, P.X., Sunindijo, R.Y. & Dainty, A.R. 2014. A mixed methods research design for bridging the gap between research and practice in construction safety. *Safety Science*, 70: 316–326.

16 Using Convergent Mixed Methods to Explore the Use of Recycled Plastics as an Aggregate for Concrete Production in South Africa

Nishani Harinarain, Robert A. Lukan, and Bevan Naidoo

Summary

In a world where the population is ever-increasing and the wealth of individuals is constantly growing, the amount of waste materials produced, such as plastics, metals and glass, is enormous. Plastics must be recycled efficiently to reduce the impact that they have on the environment. A convergent mixed methods research design was used in this study to investigate the willingness of construction industry stakeholders to use recycled plastics as a substitute for sand as an aggregate in concrete. This involved the collection of quantitative and qualitative data using surveys and interviews. Of 189 questionnaires that were distributed, 52 completed questionnaires were returned and used, whilst six interviews were conducted with built-environment professionals. This was followed by the analysis of the data collected. The results from the survey were integrated with the findings from the interviews. It was discovered that different types of strengths found in concrete would change with the substitution of recycled plastic aggregates and, therefore, professionals were willing to use it in non-load bearing building components such as concrete sidewalks, paving and roof tiles. Further research is needed to ensure the right level of substitution of plastic aggregates for sand is used, and the cost-effectiveness of such substitution.

Introduction

For human beings, plastics serve as a beneficial product in everyday life. They are versatile, lightweight, flexible and relatively cheap to make and purchase. Owing to these properties, plastics are produced on a mass scale throughout the world, and it is predicted that, by the year 2050, 1,800 million tonnes of plastic would be produced annually (Qualman, 2017).

Most plastics, such as containers and bottles, are made up of polyethylene terephthalate (PET). This is a petroleum-based plastic which does not decompose in the same way that other bio-degradable materials do. As such, these PET plastics have a lifespan of over 450 years (Harris, 2019), hence their contribution to the accumulation of plastic waste.

This build-up of plastic waste harms the environment, as it threatens marine and wildlife with over 700 marine species killed by plastic (Felt, 2019; Ritchie &

DOI: 10.1201/9781003204046-16

Roser, 2018). Plastic waste contaminates drinking water and soil by releasing toxins that seep into the soil and eventually mix with groundwater, which is used for drinking, as well as the soil, used to grow fruit and vegetables (Felt, 2019).

The increased production of plastic will result in less plastic being re-used and more plastic being wasted, filling up the landfill sites at an extraordinary pace. Unfortunately, because of the increased waste production, landfill sites are being forced to shut down because they have run out of space to hold any more waste, which only adds to this growing concern and puts more strain on remaining landfill sites (Felt, 2019).

To solve the problem of the slow decomposition of plastic, a significant investment in recycling activities is needed. Recycling plastics can result in plastic being re-used for the same or another purpose, thereby reducing the amount of plastic waste. An option for recycling in the construction industry, to reduce the pollution of plastic and the amount of waste sent to landfills, is to incorporate recycled plastic as an aggregate during concrete production.

The purpose of his study was to determine the perceptions of engineers, quantity surveyors and contractors of the use of recycled plastic as a substitute for fine aggregates during concrete production. This was done by using a convergent mixed methods research design. This involved questionnaires and interviews, followed by an analysis of the quantitative and qualitative datasets. The results from the quantitative method were integrated with the findings of the qualitative method.

Plastics and Concrete Production

Plastics have become an integral part of everyday life. The demand for plastics has risen consistently, as their unique properties make it possible to use them in various applications. It is for this reason that manufacturers willingly produce mass quantities of plastic to keep up with the large demand (Ritchie & Roser, 2018).

Recycling is essential to save energy and to ensure that the world does not deplete natural resources. This is emphasised further by the advantages of recycling, namely, the sustainability of natural resources, minimising pollution (reducing the level of greenhouse gas emissions brought about by the manufacturing process and reducing industrial or commercial waste) and saving space in landfill sites for future generations (Renewable Resources Coalition, 2019).

Concrete is one of the most used materials and an extremely vital material in the construction industry. Known for its strength and durability, concrete is used to construct foundations, slabs, columns, beams, roofs, etc. The global annual production of concrete amounts to ~10 billion tons, which is ~4.1 billion cubic metres (Global Concrete Production, 2016).

Aggregates in Concrete

Concrete is composed of cement, water, fine aggregate (sand) and coarse aggregate (gravel and rock) (Portland Cement Association, 2018). The final composite material produced is a workable substance which eventually hardens over time. Aggregates make up between 60% and 80% of a normal concrete mix and, therefore, take up most of the volume of concrete, making them a fundamental component of concrete. The size, shape and composition of the aggregates in the concrete have an overall effect on the workability, durability, strength, weight and shrinkage of the concrete (Seegebrecht, 2019).

The focus of this study was on eliciting the perceptions of stakeholders regarding the substitution of sand with recycled plastic during concrete production. South Africa accounts for about 45.5 million tons or 18.9 million cubic metres of concrete production (Muigal *et al.*, 2013) and ~52 million tons of sand were sold as a fine aggregate. If South Africa could substitute even just a percentage of sand aggregate with recycled plastic, South Africa could dramatically increase the recycling rate of plastics in the country.

The Effect of Plastic Aggregates on the Physical Properties of Concrete

Concrete has many important physical properties (as listed in Table 16.1) that persuade professionals in the construction industry to use it (Neville, 2012). Various studies have been conducted on the use of plastic aggregates in concrete production (Gu & Ozbakkaloglu, 2016) and some of these studies are presented in Table 16.1. However, there have not been many large-scale projects in South Africa where recycled plastic aggregates have been implemented. Perhaps the only use of recycled plastic that is becoming prevalent in the construction industry currently is for road construction (Businesstech, 2019). Other physical properties of concrete include shear/torsion strength, durability, fire resistance, impact resistance, thermal and acoustic insulation, and modular ratio.

Table 16.1 Physical properties of concrete and tests

Physical properties	*Description*	*Tests*
Workability	The ease with which concrete, in its fresh state, can be placed and shaped to make different components.	Gu and Ozbakkaloglu (2016) and Tang *et al.* (2008) found that, where coarse aggregates were substituted with polystyrene aggregates, the slump was similar to that of the control concrete.
Density/unit weight	The degree of compactness of the material.	Osei (2014) found that the introduction of recycled plastic aggregate resulted in a reduction in the mass of the mixture.
Compressive strength	The ability of concrete to resist loads pressing down on it.	Hameed and Ahmed (2018) indicated that the compressive strength diminished as the level of plastic increased.
Tensile (splitting) strength	The ability to resist being pulled apart.	Mohammadhosseini *et al.* (2018) show that, if recycled plastic aggregates exceed 5%, the tensile strength of concrete tends to drop lower than regular concrete.

(*Continued*)

Table 16.1 (Continued)

Physical properties	*Description*	*Tests*
Flexural strength	The ability of concrete members to resist bending.	Rai *et al.* (2012) showed that, as the amount of plastics substituted for fine aggregates was increased, the strength, both compressive and flexural, would decrease slightly.
Abrasion resistance	The physical wearing down because of hard particles on a solid surface.	Ferreira *et al.* (2012) stated that the size, toughness and shape of the plastic aggregates used will influence the abrasion resistance of concrete.
Elastic modulus	A stress to strain ratio for hardened concrete.	Choi *et al.* (2005) found that, as the percentage of PET increased in the concrete, the modulus of elasticity decreased.
Poisson's ratio/ toughness	The ratio of lateral strain to longitudinal strain.	Kou *et al.* (2009) found that the values of the Poisson's ratio increased with the increase of recycled PVC aggregate.
Water absorption/ porosity	The relationship between the strength and the total volume of voids.	Coppola *et al.* (2018) found that, as the substitution levels rose, so did the rate of water absorption.
Chloride migration	Ability to allow chloride ions to enter through concrete affecting the reinforcement.	When dealing with chloride migration in concrete containing RPA, the chloride permeability tends to be higher than that of regular concrete (Silva *et al.*, 2013).
Shrinkage	The contraction of hardened concrete due to the loss of capillary water.	Gu and Ozbakkaloglu (2016) established that, as the substitution level of plastics aggregates increased, the drying shrinkage of the concrete increased. However, Kou *et al.* (2009) and Silva *et al.* (2013) found that, when PET was used instead of EPS, the drying shrinkage was lower than that of regular concrete.

Source: Original.

Methodology

A convergent mixed methods research design was used in this study as indicated in Figure 16.1. Mixed methods research design involves the combination of methods that are both qualitative (an interpretive understanding of underlying reasons, opinions and motivations) and quantitative (which harnesses natural sciences, in order to build numerical data and state facts) to allow for complete and in-depth research. A mixed methods approach enables the researchers to make better use of the advantages of both methods (Panke, 2018), thereby effectively neutralising the effect of their disadvantages. This also adds to the reliability and validity of the research project (Creswell & Creswell, 2014).

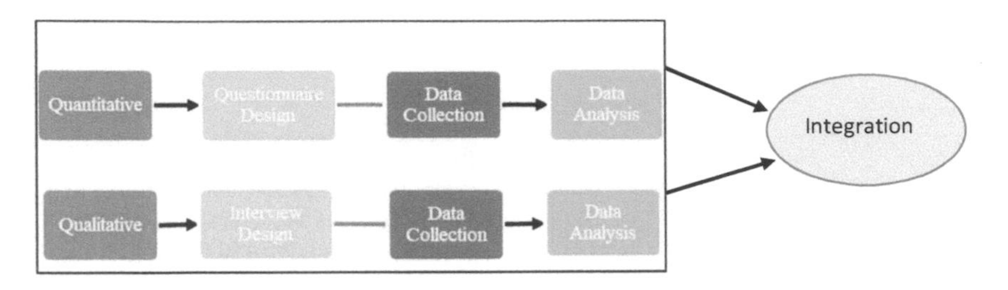

Figure 16.1 Convergent mixed methods design was used in this study

Table 16.2 Sample size of various consulting firms in South Africa

	Consulting firms in SA			
	Engineers	*Quantity surveyors*	*Civil contractors*	*Total*
Firms in South Africa	810	702	349	1,861
Firms in KwaZulu-Natal	145	115	73	333
Firms in Durban	64	73	52	189

Source: Original.

This study incorporated a combination of both questionnaires (quantitative) and semi-structured interviews (qualitative) in eliciting the perspectives of various role-players in the construction industry regarding the subject.

One of the tranches of this convergent mixed method research design was the use of questionnaires. These questionnaires were distributed to engineers, contractors and quantity surveyors, as these professionals were more likely to have the most experience in dealing with concrete, in order to obtain their opinion on the topic of recycled plastic aggregates in concrete.

Various databases, such as Consulting Engineers of South Africa (CESA), The Association for South African Quantity Surveyors (ASAQS) and the South African Forum of Civil Engineering Contractors (SAFCEC), were used to determine the sample size for this study. The total population comprised 1861 consulting firms. Cluster sampling, which is a form of probability sampling, was used to select the participants because the sample was limited to firms within Durban, KwaZulu-Natal. Cluster sampling was used to refine the search from the firms in the nine provinces in South Africa to those in KwaZulu-Natal Province, and then further to firms in Durban. The final sample size was 189, as shown in Table 16.2. Emails were sent to prospective participants. In the email, what the research was about was explained and an informed letter of consent was attached in order to ensure good ethical practice.

The qualitative data were collected through semi-structured, online interviews. Interviews were deemed to be appropriate because in-depth data could be gathered from a small group of practising professionals within the construction industry and they provided a deep and much more substantial response (Creswell & Creswell, 2014).

For the qualitative research component, purposive sampling was used to select the interviewees.

Semi-structured, online interviews were conducted with six professionals. While the number of interviewees for a study is often predicated on the attainment of data saturation, Galvin (2015) stated that a common sample size for qualitative research falls within the range of 5–10 in-depth interviews. The six interviews conducted were deemed to be appropriate because each interviewee had over 15 years of experience. Each interview lasted ~45 minutes.

A total of 52 questionnaires were completed and returned. According to Fincham (2008), an acceptable response rate for a survey, that is sent out to a large group of potential participants via email, ranges between 25% and 30%. Therefore, the response rate of 27.5% achieved for this study was deemed to be acceptable. The completed questionnaires were captured on SPSS for analysis. To ensure the reliability of quantitative data, Cronbach's alpha was used. Cronbach's alpha is a value which is expressed between 0 and 1. A minimum Cronbach's alpha of 0.7 was required to ensure the reliability of the data obtained (Tavakol & Dennick, 2011). The mean and the standard deviation were used to interpret the data.

The interviews were transcribed and uploaded into the NVivo software for analysis. The data were then analysed using thematic analysis. Thematic analysis is a form of analysis that is used to examine and record patterns within the data that were gathered (Creswell & Creswell, 2014).

In the aftermath of data analysis, the results were integrated in a manner resembling triangulation. Triangulation is the process of combining two different methods to study a particular concept. This is an important strategy to use for any mixed methods research project because it is a vital component which reduces the possibility of bias occurring within the research (Mertens & Hesse-Biber, 2012). Triangulation enables researchers to analyse and review data to ascertain how they either complement or disagree with each other. Researchers can draw further conclusions from the data, based on the findings obtained from the triangulation process.

Results and Discussion

Quantitative Data Analysis

Engineers formed the majority of the participants, followed by the quantity surveyors. The respondents comprised 76% male, and 46% of the respondents had over 16 years' experience, as indicated in Table 16.3.

All the participants thought it was very important to improve recycling practices to achieve sustainability. Most (80%) of the respondents thought that it was very important to reduce the amount of waste in landfills, because of the adverse effects on the environment. The vast majority (96%) of the respondents thought it was very important to recycle in order to preserve natural resources.

A majority (55%) of the respondents experienced a shortage of sand or stone or both during a project on which they had worked, which resulted in an increase in the prices of these two natural resources. The shortage of sand and stone could

Table 16.3 Demographics, professions and experience of participants

Gender	Male	76%
	Female	24%
Profession	Engineers	35.29%
	Quantity surveyors	33.33%
	Contractors	15.69%
	Construction project managers	15.69%
Practising experience (years)	0–5	27%
	6–10	22%
	11–15	10%
	16–20	8%
	21–25	8%
	25<	25%

Source: Original.

be because of a number of pressing issues, such as supplier shortages, over-priced aggregate materials and legislative issues. However, it was found in this study that the main reason was because of the availability of the resources only in remote locations, which led to logistical issues. This observation signalled a demand for an alternative material that can replace sand and/or stone as naturally-occurring aggregates in concrete.

Applying the Concept of Recycled Plastic in Concrete

Most (90%) of the participants responded that plastic is an important material that must be recycled, confirming that new innovations are needed to increase the recycling rate of plastics around the world. A majority (58%) of the respondents had previously heard of the use of recycled plastic aggregates in concrete but only four respondents had personally experienced the use of recycled plastic in brick production for low-cost housing and recycled plastic in concrete to improve the strength of beams.

Possible Applications of Concrete with Recycled Plastic Aggregate

Participants in the research were asked to use their relevant experience and knowledge of concrete to determine the possible applications of recycled plastic in concrete. The Cronbach's alpha for the questions based on the application of recycled plastic aggregates in concrete was 0.74, hence regarded as being acceptable (Tavakol & Dennick, 2011). Each application of recycled plastic in concrete is listed and ranked in Table 16.4, according to the mean scores obtained from the responses. The researchers found that there were three major applications, namely, concrete sidewalks, concrete asphalt/paving and concrete roof tiles, in which the respondents believed that the substitution of recycled plastic aggregates for sand could be highly likely. One possible reason for this choice was related to these applications being concrete items that are not load-bearing, or which do not experience considerable stress. These applications are extremely common and are used in large quantities, which means that more plastic can be recycled and used in these components, leading to an increased rate of recycling.

The respondents did not think that recycled plastic should be used in concrete foundations (strip footings, bases etc.), reinforced concrete (stairs, columns, beams etc.) and structural concrete for bridges and large buildings. A possible reason for this was that the respondents believed that these applications are an integral part of any building and that, as professionals, they would not risk replacing these with recycled plastic, to forestall failure.

Effects of Recycled Plastic Aggregates on the Physical Properties of Concrete

Participants were asked to apply their experience in estimating the effect of recycled plastic aggregate on the physical properties of concrete (as shown in Table 16.5). This dataset returned a Cronbach's alpha score of 0.715 which was acceptable (Tavakol & Dennick, 2011). The participants believed that the compressive strength was the property that would be most affected and/or altered when incorporating recycled plastic aggregates into concrete. The other properties that the participants thought would be affected included: flexural strength; tensile strength; unit weight/density; and concrete grades. The physical property that was thought to be affected the least by the introduction of recycled plastic aggregates was abrasion resistance.

Willingness of Construction Industry Stakeholders to Incorporate Use of Recycled Plastics as Aggregates in Concrete

It was found that 75% of the respondents believed that it was more beneficial to implement recycling regardless of its cost, while 25% felt that, because of how expensive recycling can be, its benefits were simply not worth it. However, despite some professionals rejecting recycling because of its costs, 96% of these professionals

Table 16.4 Uses of concrete with plastic aggregate

Application	*Mean*	*Std dev.*	*Rank*
Concrete sidewalks	4.49	0.73	1
Concrete asphalt/paving	4.33	0.93	2
Concrete roof tiles	4.22	0.92	3
Concrete blocks for residential/commercial use	4.00	0.98	4
Mortar mix for binding bricks	3.59	1.08	5
Concrete surface beds and slabs	3.47	1.06	6
Precast/pre-stressed concrete	3.22	1.32	7
Concrete foundations (strip footings, bases etc.)	3.27	1.23	8
Reinforced concrete (stairs, columns, beams etc.)	2.96	1.20	9
Structural concrete for bridges and large buildings	2.61	1.28	10

Source: Original.

were willing to use plastic aggregates in concrete in the near future. As a result of this, 92% of them agreed that the rate of recycling in South Africa would be increased if recycled plastic aggregates were used in concrete and the consumption of natural aggregates, such as sand, could be reduced.

Qualitative Data Analysis

Thematic content analysis is used to identify common and repeating patterns amongst the qualitative data and was used to analyse the interviews. Table 16.6 shows the demographic information of the interviewees. Engineers formed the majority of the participants. The interviewees comprised 83% male, and 50% of the respondents had over 20 years' experience.

An overview of the most important themes from all the interviews is shown in Figure 16.2, namely: the effects on the physical properties of concrete, possible concrete applications and the opinions of the stakeholders.

Table 16.5 Results for effect on physical properties of concrete

Physical property	*Mean*	*Std dev.*	*Rank*
Compressive strength	3.65	0.84	1
Flexural strength	3.61	0.92	2
Tensile strength	3.59	0.94	3
Unit weight/density	3.45	0.94	4
Concrete grades	3.41	0.88	5
Durability	3.33	0.93	6
Fire resistance	3.25	0.88	7
Shrinkage	3.24	1.14	8
Slump	3.16	1.10	9
Workability	3.12	1.05	10
Water absorption	3.04	1.26	11
Chloride migration	2.96	0.98	12
Poisson's ratio	2.94	1.08	13
Abrasion resistance	2.73	1.23	14

Source: Original.

Table 16.6 Demographics, profession and experience of the interviewees

Gender	Male	83%
	Female	17%
Profession	Engineers	66%
	Quantity surveyors	17%.
	Construction project managers	17%
Practicing experience (years)	15–20	33%
	21–25	50%
	25<	17%

Source: Original.

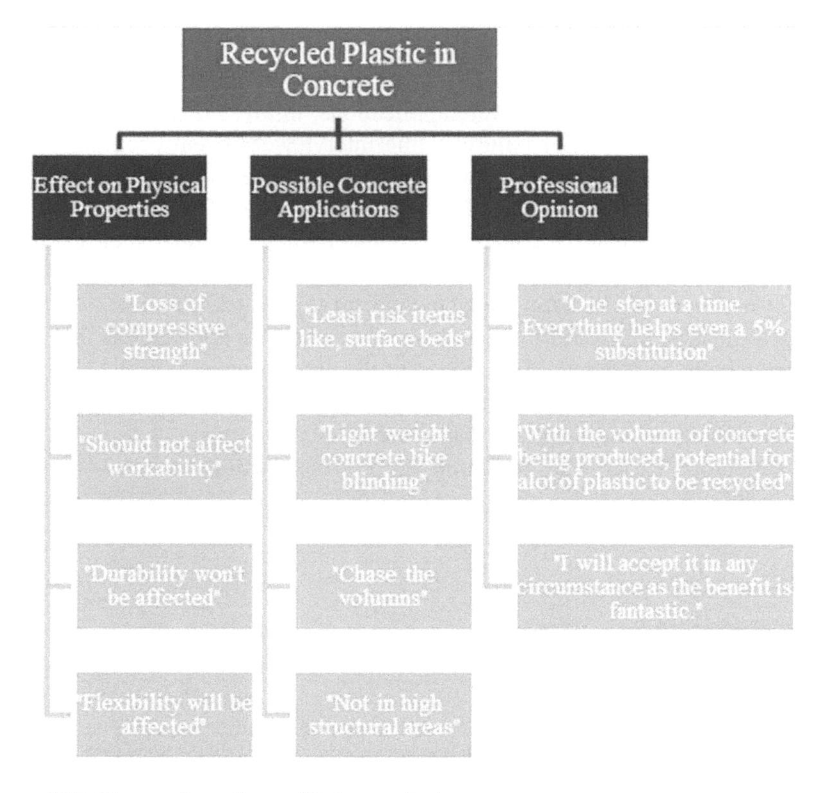

Figure 16.2 Themes from the qualitative analysis

Effects of Recycled Plastic on the Physical Properties of Concrete

Most of the interviewees agreed that, in general, most of the physical properties would be affected in one way or another. One interviewee stated that "not all change is bad change" meaning that, by using recycled plastic as an aggregate, some physical properties might actually improve based on the expected use or need for that concrete. Most interviewees (70%) believed that flexural and tensile strength would be improved by introducing recycled plastic aggregates. This, in turn, could produce a concrete solution that would be preferred for applications that require a higher flexural or tensile strength. On the other hand, at least 50% of the interviewees agreed that a concrete mix with recycled plastic aggregates would have a reduced compressive strength "due to the flexibility of plastic". Compressive strength is considered to be a very important physical property of concrete, which is a cause for concern. All interviewees agreed that the chemical physical properties of concrete, such as fire resistance, chloride mitigation and water absorption were highly unlikely to be affected.

Possible Applications for Concrete

Most interviewees were of the opinion that concrete with recycled plastic aggregates should be used in low-risk concrete applications, in areas of a building where it would have the least risk of failure. Some interviewees even recommended that concrete with recycled plastic aggregates should be used exclusively in blinding as it is a concrete application that has a low risk of failure. The issue with this suggestion is that the amount of concrete used for blinding is a fraction of that used in an entire project. It is because of this that a few interviewees suggested that, when considering concrete containing recycled plastic aggregates, one should "chase the volumes", that is, this type of concrete should be used in applications that require large quantities of concrete. The aim is to use large amounts of recycled plastic waste in an effort to solve the problem of plastic pollution.

Also, the interviewees agreed that concrete containing recycled plastic aggregates should not be used for structural applications, such as beams, columns and bridges, because these applications are high-risk areas of a building or structure.

Opinions of Practicing Professionals

The final theme discussed was whether the interviewees within the industry would be willing to use recycled plastic aggregates in the near future. Some interviewees said that they would accept using this type of concrete under "any circumstances", as the benefits of this would be worth the sacrifice of other aspects, such as cost. These interviewees believed that the huge problem of plastic pollution could be solved. Other interviewees would only accept using concrete with recycled plastic aggregates if it would be a cost-effective solution as well as satisfying every test to ensure that it is a safe solution.

The interviewees believed that the recycling rate of plastic would increase if concrete with recycled plastic aggregates were used or specified through regulations in an effort to ensure concrete suppliers incorporate recycled plastics as aggregates. Another interviewee explained that concrete is always produced in large quantities and, thus, there is a huge amount of potential for the recycling rate of plastic to increase if recycled plastic aggregates were incorporated into concrete.

Several interviewees believed that this was an extremely interesting idea with great potential use in the industry. However, they did feel that more research should be done in terms of feasibility and viability when using recycled plastics as aggregates. One interviewee believed that more prototypes or pilot projects incorporating recycled plastics were needed to gain more data on the topic. Another interviewee stated that producers and suppliers of building materials should be pioneering this revolution of aggregates and should take the initiative on this issue.

Integration

The researchers found that there was consensus amongst the responses elicited from both the questionnaires and the interviews.

Participants in the questionnaires and interviews had very similar opinions with regard to the possible applications of concrete with recycled plastic aggregates and the effects that recycled plastic would have on the physical properties of concrete. The results from the interviews reinforced what was found in the results of the questionnaires with regard to these concepts. The participants agreed that flexural and tensile strength were most likely to improve with recycled plastic aggregates and compressive strength and unit weight/density would be most likely to be negatively compromised by recycled plastic aggregates.

The responses from both the questionnaires and interviews revealed that the best applications for concrete consisting of recycled plastic aggregates would be non-load-bearing structures with a low risk of failure such as concrete sidewalks, concrete asphalt/paving and concrete roof tiles and blinding. Similarly, most participants agreed that concrete with recycled plastic aggregates should certainly not be used for structural concrete in bridges and large buildings, reinforced concrete and structural components such as columns and beams because of the high structural component and high risk of failure.

The majority of the participants in the questionnaires and interviews agreed that the concept of using recycled plastic as an aggregate in concrete production was possible and could become a viable option and were willing to use concrete containing recycled plastic aggregates in a future project, which would increase the rate of recycling in South Africa.

Most participants also stated that they would specify or implement this concept in future projects after considering its benefits. However, some participants were of the opinion that because of the high competition for top prices in the industry, the concept would have to prove to be the more cost-effective option. It was suggested further that, if this concept proves to be beneficial, perhaps the government should implement legislation in order to make it a legal requirement to replace a certain percentage of natural aggregates with recycled plastic in all future projects.

Implications for Practice and Research

This study is important because it contributes towards deepening knowledge relating to the utility of recycled plastics in the construction industry. Plastic recycling has gained importance as a result of the global recognition of the negative impacts of plastic pollution and the importance of reducing its ecological footprint. The construction industry will benefit from this study because it has considerable potential to contribute to the overall recycling rate of plastic, and the study raises awareness of this possibility which, hopefully will, in turn, benefit the waste management and pollution control processes in the country.

The use of a convergent mixed methods design enabled the researchers to obtain a broader, richer, perspective on the topic and increased the credibility of the study. The researchers found that the convergent mixed methods research design greatly assisted in collecting the data quickly, as the data collection occurred concurrently. Integrating the data provided insights that would not have been obtained had only one method been used. The low response rate experienced when collecting quantitative data posed a challenge. However, the researchers overcame this by sending reminders to the respondents and structuring the questionnaire in such a way that made it easy to comprehend and answer. Researchers who wish to conduct convergent mixed methods research design in the future are advised to spend more time on planning their questionnaire and interview schedules to improve the response rate and assist in data collection and analysis. It is encouraging to see that more researchers are heeding the call to use various approaches.

Conclusion

Reducing plastic waste serves as a beneficial yet difficult challenge to overcome, largely because of the decomposition rate as well as the ever-increasing production rate. The build-up of plastic waste has affected the environment adversely. Although there has been a recent transition towards the recycling of plastic waste, the incidence of such waste persists. However, the use of plastic waste as aggregates in concrete in the construction industry might provide a suitable option to reduce the amount of plastic waste in the environment.

A convergent mixed methods design was adopted to achieve the aim of the study. The use of this approach produced conclusions that results from a single method might not have produced.

The findings suggested that the physical properties in concrete would be subject to change (negative or positive), with the substitution of recycled plastic aggregates for sand. It is recommended that the potential use of recycled plastic aggregates in concrete would be in components that are made up of high volumes of concrete that are not structural or load-bearing. This is essential to ensure that the use of recycled plastic aggregates does not lead to the failure of a building or structure while incorporating more recycled plastic in high volumes of concrete. It can be concluded that professionals in the South African Construction Industry are most willing to use recycled plastics as aggregates in concrete. Therefore, it is recommended that further research and tests be done to ensure that the concrete made with recycled plastic aggregates will be reliable for use as a substitute aggregate.

References

Businesstech. 2019. *Government Wants to Use Waste Plastic in South Africa's Roads.* [Online]. Available at: https://businesstech.co.za/news/technology/341915/government-wants-to-use-waste-plastic-in-south-africas-roads/ [Accessed: 26 September 2019].

Choi, Y., Moon, D., Chung, J. & Cho, S. 2005. Effects of waste PET bottles aggregate on the properties of concrete. *Cement and Concrete Research*, 35(4): 776–781.

Coppola, B., Courard, L., Michel, F., Incarnato, L., Scarfato, P. & Di Maio, L. 2018). Hygrothermal and durability properties of a lightweight mortar made with foamed plastic waste aggregates. *Construction Building Materials*, 170: 200–206. https://doi.org/10.1016/j.conbuildmat.2018.03.083

Creswell, J. & Creswell, J. 2014. *Research Design: Qualitative, Quantitative and Mixed Method Approaches*. 4th edition. SAGE, Los Angeles, CA.

Felt, C. 2019. *Plastic Waste: Environmental Effects of Plastic Pollution*. [Online]. Thriveglobal.com. Available at: https://thriveglobal.com/stories/plastic-waste-environmental-effects-of-plastic-pollution/ [Accessed: 15 April 2019].

Ferreira, L., de Brito, J. & Saikia, N. 2012. Influence of curing conditions on the mechanical performance of concrete containing recycled plastic aggregate. *Construction and Building Materials*, 36: 196–204.

Fincham, J. 2008. Response rates and responsiveness for surveys, standards, and the journal. *American Journal of Pharmaceutical Education*, 72(2): 43.

Galvin, R. 2015. How many interviews are enough? Do qualitative interviews in building energy consumption research produce reliable knowledge? *Journal of Building Engineering*, 1: 2–12.

Global Concrete Production. 2016. *California: Silicon Solutions*. [ebook] Available at: http://siliconesolutions.com/media/pdf/gcp.pdf [Accessed: 30 January 2019].

Gu, L. & Ozbakkaloglu, T. 2016. Use of recycled plastics in concrete: A critical review. *Waste Management*, 51: 19–42.

Hameed, A. & Ahmed, B. 2018. Employment the plastic waste to produce the lightweight concrete. *Energy Procedia*, 157: 30–38.

Harris, W. 2019. *How Long Does It Take for Plastics to Biodegrade?* [Online]. HowStuffWorks. Available at: https://science.howstuffworks.com/science-vs-myth/everyday-myths/how-long-does-it-take-for-plastics-to-biodegrade.htm [Accessed: 28 January 2019].

Kou, S., Lee, G., Poon, C. & Lai, W. 2009. Properties of lightweight aggregate concrete prepared with PVC granules derived from scraped PVC pipes. *Waste Management*, 29(2): 621–628.

Mertens, D. & Hesse-Biber, S. 2012. Triangulation and mixed methods research. *Journal of Mixed Methods Research*, 6(2): 75–79.

Mohammadhosseini, H., Tahir, M. & Sam, A. 2018. The feasibility of improving impact resistance and strength properties of sustainable concrete composites by adding waste metalized plastic fibres. *Construction and Building Materials*, 169: 223–236.

Muigal, R. Alexandar, M.G. & Moyo, P. 2013. Cradle-to-gate environmental impacts of the concrete industry in South Africa. *Journal of South African Institute of Civil Engineering*, 55(2): 2–7.

Neville, A. 2012. *Properties of Concrete*. 5th edition. Pearson Education, Harlow.

Osei, D. 2014. Experimental investigation on recycled plastics as aggregate in concrete. *International Journal of Structural and Civil Engineering Research*, 3(2): 168–174.

Panke, D. 2018. *Research Design and Method Selection*. SAGE Publications, London.

Portland Cement Association. 2018. *Aggregates*. [Online]. Available at: https://www.cement.org/cement-concrete-applications/concrete-materials/aggregates [Accessed: 23 April 2019].

Qualman, D. 2017. *Global Plastics Production, 1917 to 2050 » Darrin Qualman*. [Online]. Available at: https://www.darrinqualman.com/global-plastics-production/ [Accessed: 28 January 2019].

Rai, B., Rushad, S., Kr, B. & Duggal, S. 2012. Study of waste plastic mix concrete with plasticizer. *ISRN Civil Engineering*, 2012: 1–5.

Renewable Resources Coalition. 2019. *Recycling Advantages & Disadvantages: The Ups & Downs of Recycling*. [Online]. Available at: https://www.renewableresourcescoalition.org/recycling-advantages-disadvantages/ [Accessed: 5 March 2019].

Ritchie, H. & Roser, M. 2018. *Plastic Pollution*. [Online] Our World in Data. Available at: https://ourworldindata.org/plastic-pollution [Accessed: 28 January 2019].

Seegebrecht, G. 2019. *Aggregate in Concrete - The Concrete Network*. [Online]. Concrete network. Available at: https://www.concretenetwork.com/aggregate/ [Accessed: 23 April. 2019].

Silva, R., de Brito, J. & Saikia, N. 2013. Influence of curing conditions on the durability-related performance of concrete made with selected plastic waste aggregates. *Cement and Concrete Composites*, 35(1): 23–31.

Tang, W., Lo, Y. & Nadeem, A. 2008. Mechanical and drying shrinkage properties of structural-graded polystyrene aggregate concrete. *Cement and Concrete Composites,* 30(5): 403–409.

Tavakol, M. & Dennick, R. 2011. Making sense of Cronbach's alpha. *International Journal of Medical Education*, 2: 53–55.

17 Adaptive Mixed Methods Research for Evaluating Community Resilience and the Built Environment

Sandra Carrasco and Temitope Egbelakin

Summary

In this chapter, adaptive mixed methods research approaches are explored for examining community resilience and its implications for creating a safer built environment. Mixed methods research design, comprising two case studies in the Philippines and New Zealand, are compared and integrated in this chapter. In the cases presented, mixed methods research design was used as a core methodological approach and incorporated multiple instruments for data collection and analysis. In the Philippines case, the motivations of post-disaster housing reconstruction beneficiaries to perform housing adaptations were examined. In the New Zealand case, motivations of building owners and behaviours to adopt earthquake retrofitting in their vulnerable buildings were analysed. The findings presented in this chapter illustrate a series of subsequent variations in the data collection instruments and research strategies that were necessary to address the objectives of the studies in contexts of permanent change because of the ongoing disaster recovery. The adaptability required in research about community resilience to disaster provides opportunities to incorporate novel approaches and techniques in the research process. However, in this study, it is stressed that maintaining consistency across the data collection instruments and analysis is paramount to ensure high-quality data and rigorous research. Consistency in research will contribute to meaningful research-informed impacts to enable decision-making to achieve resilient communities and build environments.

Introduction

Research about the complexities and multiple dimensions of the causes of disasters and their impacts on the interactions between people and their built environment requires an inter-disciplinary approach that links social sciences, natural sciences, and engineering, combining diverse research strategies (Faber *et al.*, 2014; Gilligan, 2021; Taylor *et al.*, 2014). Riazi and Candlin (2014) pointed out that it is necessary for decision-making about the methodology to respond to "what works" and what is appropriate to address pertinent research questions and the context in which they are being asked. Similarly, Johnson *et al.* (2007) called for redefining research

DOI: 10.1201/9781003204046-17

approaches towards the construction of a "workable" definition that integrates the paradigms of qualitative and quantitative research rather than seeing them as a mutually exclusive binary choice. Furthermore, in recent studies conducted in the wake of the COVID-19 pandemic, the importance of adaptability in research approaches was re-examined to address the uncertainties in the context of this health crisis (Bueddefeld *et al.*, 2021; McEachan *et al.*, 2020; Tavares *et al.*, 2021).

Therefore, addressing the intricacies in research about disaster and the interactions between people and built environment requires diverse research methodologies, which implies a combination of different approaches that researchers should choose based on the nature of the research and flexibility according to the research objectives and the researchers' philosophical preferences (Amaratunga *et al.*, 2002). Therefore, in this chapter, the need for adaptability of mixed methods research designs for investigating community resilience and interactions with the built environment is addressed to create safer urban environments. The chapter is focused on the implementation of various data collection instruments and approaches in two international case studies, in the Philippines and New Zealand, to analyse the research methodologies applied in these studies, reflecting the need for adaptability in mixed methods and methodologies of research about disaster, to address the permanently changing conditions of disaster recovery and preparedness.

The focus of the case studies was on examining the behaviours of communities and how these influence their decision-making about their properties and implications for creating a safer built environment.

The trends in research about disasters and the built environment are discussed in the following section. Then, the methodology adopted in this research is explained in the third section. The focus of the fourth section is on the approaches to adaptability in researching community resilience and disaster in the built environment. To do so, the methodologies adopted in the two case studies are analysed. The main issues and advantages of both cases are discussed in the fifth section and the implications for practice and research are analysed. Finally, conclusions are drawn from the empirical evidence and discussions.

Mixed Methods Research in Studies about Disaster and the Built Environment

Methodologies used in research about disasters provide different insights into the strategies used and whether they focus on either quantitative or qualitative research or integrate both. For instance, Suárez *et al.* (2020) analysed academic articles in which urban resilience was measured and observed that quantitative methods were used in more than 76%, while qualitative methods were used in 16% and mixed methods in 7%. Witt and Lill (2018) conducted a systematic review of 156 articles about resilience to disaster in the built environment and found that qualitative research methods were used in 66%, quantitative methods in 27% and mixed methods research was used in only 7%.

The use of mixed methods design in research about disaster and the built environment has emerged only in the last 15 years, as shown in Figure 17.1, based

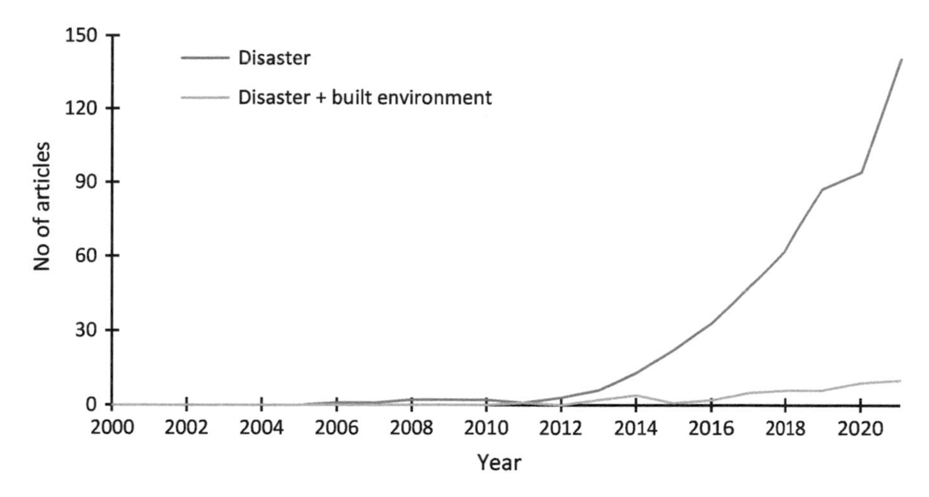

Figure 17.1 Research articles about disaster and built environment using mixed method research during 2000–2021

on the number of publications during 2000–2021 in the Thomson Reuters Web of Science. Figure 17.1 shows the authors' compilation based on two searches that were done using the Web of Science. The first search term used was "mixed method" in the topic search fields of "disaster" and "disaster resilience". The second search term used was "mixed method" in the topic and search fields of "disaster", "disaster resilience" and "built environment". Researchers made a careful distinction in each of the searches to confirm the accuracy of the results. Special considerations were given to variations of "mixed methods" such as "mixed-methods", "mixed research methods", "mixed methodology", and "qualitative and quantitative methods".

Conversely, use of mixed methods in health studies has increased rapidly since the approach was first adopted in the 1990s (Doyle *et al.*, 2009). Additionally, use of mixed methods in social and behavioural studies increase rapidly after 2006 (Timans *et al.*, 2019). Mixed methods research has been used progressively in various areas of social science in which it had its origins rather than in the "pure disciplines" (Tashakkori & Teddlie, 2010). Furthermore, the acceptance of mixed methods research expanded rapidly in inter-disciplinary studies in which complex problems were addressed in contrast to the straightforward approaches of traditional disciplines (Timans *et al.*, 2019).

The complex issues evaluated in research about disaster are being addressed increasingly through mixed methods research. In the years after 2015, especially, the use of mixed methods in research about disaster has gained momentum as it increased by 536% in 2021. Even though the use of a mixed methods approach in research about disaster is increasing rapidly, the trends in research about disaster in the built environment are incipient. The approach represents only 9% of the research studies about disaster conducted until 2021, although the trend is growing rapidly.

Another relevant trend is the preference of researchers for applying case studies in research design, which reflects the need for the contextualisation and empirical validation of the results in research about disaster (Andharia, 2020; Rodríguez *et al.*, 2007; Suárez *et al.*, 2020; Witt & Lill, 2018). Suárez *et al.* (2020) and Witt and Lill (2018) found that either single or multiple case studies were used in 40%–60% of the studies about urban resilience and resilience to disaster in the built environment. The adaptability in mixed methods research helps to minimise research weaknesses, enhance rigour and validity of the research, design a particular sequence, link case study design with other relevant research designs, or incorporate a particular approach or technologies into the research design (Tight, 2017).

Research Methodology

Methodological Choice

The research methodology and the variations in the research design and strategies used in both case studies are analysed separately, addressing the most relevant methodological concerns in three sub-sections: (1) Study setting, including the research context and objectives; (2) Data collection, in which the study designs, data collection instruments and the data collection approaches are analysed; and (3) Data analysis and integration in which the approaches for analysing the qualitative and quantitative data collected are discussed. Additionally, this sub-section includes the approaches used to integrate quantitative and qualitative data into relevant elements for discussion in each study.

The analysis in this chapter is focused on the research designs, including the approaches, data collection instruments and temporalities, data analysis, interpretation, and integration of the case studies. Subsequently, the findings of this study are based on the examination of convergences and divergences of both case studies (Creswell & Clark, 2017).

Case Selection and Contexts

The behaviours of communities associated with their decisions to adapt to their built environment, including their homes and other properties were analysed in both case studies. The study in the Philippines was focused on resident-driven improvements of post-disaster housing, and in the New Zealand case, the decision-making of property owners to improve existing properties to reduce risks of vulnerability to disaster was analysed.

Community Resilience and Adaptive Methodologies in Research about Disaster

In this section, research about community resilience in two cases is presented, which is crucial in order to understand the impacts of people's decision-making to create safer built environments.

Case Study 1: Self-Help Post-Disaster Housing Improvement in the Southern Philippines

Study Setting

In December 2011, the city of Cagayan de Oro, in the southern region of Mindanao, was hit by Typhoon Washi that caused massive destruction and resulted in the displacement of almost 40% of the residents. The post-disaster recovery was focused on permanently relocating the communities previously living in informal settlements along the riverbanks (Carrasco *et al.*, 2016a). These communities were moved to new settlements located 8–20 km from their pre-disaster homes.

The new houses built by the government and other organisations followed minimum standards for emergency housing in the Philippines that omitted the multiple needs of the relocated families. Since the residents received their houses, they started a process of self-help incremental housing construction. The active role of people in the co-production of their housing reflected their capacity to adapt their environment to their needs. However, it was necessary to explore the implications of the process and whether the communities had achieved resilience, avoided creating new risks, and what the impact on the development of their houses and neighbourhoods had been (Carrasco *et al.*, 2016b).

This longitudinal research was focused on a long-term analysis of the outcomes of post-disaster housing reconstruction. The residents' motivations, challenges, and implications of the incremental self-help extensions for relocation sites of original government-built houses were investigated in the study (Carrasco *et al.*, 2016a, 2017). Mixed methods research was adopted to combine the multiple perspectives and practices needed as the research context changed, such as the physical and socio-economic conditions of the settlements and communities. To capture and analyse the issues in the sites, an initial exploratory study was conducted followed by subsequent convergent research design to combine qualitative and quantitative dimensions of the study. Therefore, a bricolage approach was used in the research that consisted of a series of adjustments whilst incorporating new strategies and data collection approaches as the research developed and enabled the researchers to select the most appropriate method in each stage of the research (Sharp, 2019). Methodological bricolage, as defined by Bueddefeld *et al.* (2021), McSweeney and Faust (2019) and Denzin and Lincoln (1994) is the combination of multiple, available and accessible "practices and perspectives", such as data collection instruments and approaches, and adjusting them to address the changing local conditions throughout the different stages of the research.

Data Collection

The data was collected in three stages in 2014, 2019 and 2021. Qualitative and quantitative data were considered in the research and multiple data collection tools were used that were adjusted throughout the development of the longitudinal research (see Figure 17.2).

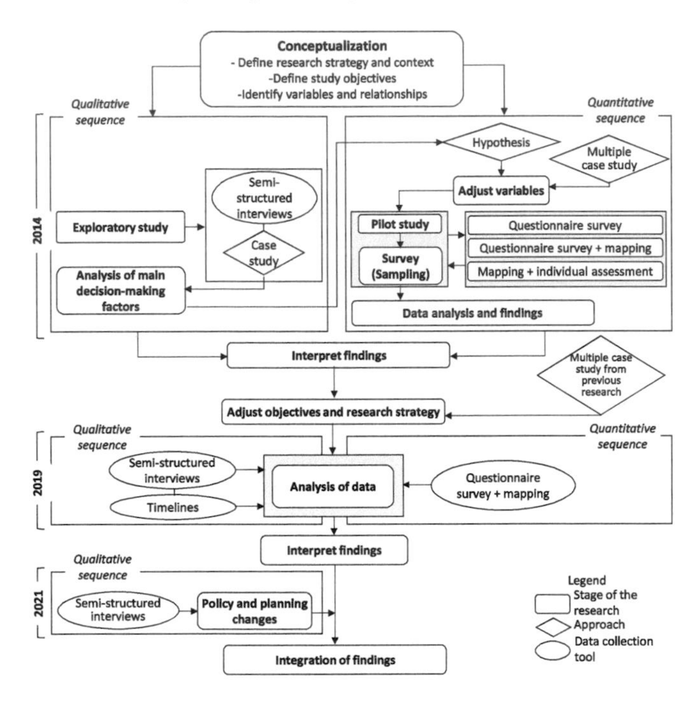

Figure 17.2 Longitudinal research process case 1

- Stage 1

  The data was collected between July and August 2014, and both qualitative and quantitative phases were conducted sequentially. The focus of the qualitative phase was on understanding the local context and the process of post-disaster housing recovery from a policy and planning perspective. Therefore, 14 semi-structured interviews were conducted with government officials, managers of NGOs, and other organisations involved. Crucial information was obtained from the interviews and initial visits to the sites, to understand the specific local conditions which were used to adjust the questionnaire survey and conduct a pilot test for the quantitative research. The quantitative phase included a pilot test using a preliminary questionnaire survey of ten families that provided insights into culturally appropriate considerations, which led to the modification of the strategies and resources to conduct the household survey. The questionnaires were administered in person by local facilitators, as proficiency in the local language was necessary, and literacy in some of the targeted households was limited.

Additionally, the surveyed families faced trauma resulting from the disaster, thus the ethical concerns included on-site support in case of distress. The focus of the quantitative component of the research was on understanding the residential satisfaction of relocated families, their living conditions, and the assessment of the safety and quality of self-built housing extensions. To do so, a questionnaire survey was administered to 254 residents, with a series of closed multiple-choice questions. The questionnaire was divided into four sections: general information, pre-disaster conditions, current conditions, risk awareness, and construction quality and safety. The last section was an integrated assessment made by the researchers to observe the wind and seismic performance of the self-built extensions. Finally, researchers mapped settlement conditions using architectural research tactics including architectural drawings, photographic surveying, and physical trace analysis (Groat & Wang, 2013; Yin, 2013; Zeisel, 1984).

- Stage 2

  The data was collected between November and December 2019 on the same research sites where the first study was conducted in 2014. Based on the findings of the first stage of the research, the objectives and research strategy were adjusted. The qualitative and quantitative phases were conducted simultaneously for this study using a convergent research approach, as shown in Figure 17.2. The quantitative research included a questionnaire survey administered to 118 resettled residents and addressed the four sections included in Stage 1. Researchers also conducted housing and settlement mapping to observe the morphological and functional changes in the sites. The qualitative component of the research included 57 semi-structured interviews in which two sections were integrated: analysis of residents' perceptions about their living conditions, and motivations for building extensions. During the interviews, the process of incremental housing construction was also considered. The findings of the 2014 stage of the research revealed difficulties in linking the housing improvements over time and the changes in the residents' socio-economic conditions. Therefore, for the 2019 stage, three timelines were considered during the interviews to address changes in household makeup, financial situations, and housing extensions built.

- Stage 3

  In November 2021, six semi-structured interviews were conducted remotely using video calls with officials from the Cagayan de Oro local government that was closely involved in projects supporting resettled communities. Since the beginning of the post-disaster recovery process, policies and initiatives to formalise tenure for residents remained unchanged and resident-built extensions were discouraged. Recently, transferring tenure to residents started and the construction of extensions is considered in the planning strategies.

Data Analysis and Integration

- Stage 1

  Two main components were combined in the analysis of the data collected during the 2014 fieldwork: the understanding of the post-disaster recovery planning and process and the multiple stakeholders involved. The analysis included the integration of the information collected during the interviews conducted and the reports of the government, NGOs and international agencies, using NVIVO software. The outcomes showed evidence of the successes and failures in the post-disaster recovery process in Cagayan de Oro and the levels of participation of the stakeholders involved, including the resettled communities. The second component was focused on the resettled communities. By reviewing the pre- and post-disaster living conditions and the satisfaction of the residents in their new houses, the factors that influenced or triggered the construction of housing extensions were analysed. Furthermore, the socio-economic conditions were contrasted with the motivations of residents to build extensions, and the current state of the settlement mapped during the fieldwork (Figure 17.3), by combining the use of the MS Excel, SPSS and AutoCAD software. Finally, the safety assessment and quality of the resident-built extensions and the analysis of the changes in the settlement were linked in the analysis.
- Stage 2

  The analysis of the data collected during the 2019 fieldwork included a comparison with the 2014 study. Therefore, the analysis of the information was carried out following the same concerns. However, the 2014 study revealed limitations to a proper understanding of the behavioural changes of the residents that motivated the construction of housing extensions. Crucial issues included changes in household structure and size, changes in their financial situation, how much families invested and the source of these investments. Following the methodology to capture the incremental housing process, developed by the Special Interest Group in Urban Settlement, Massachusetts Institute of Technology (Gattoni *et al.*, 2011; MIT, 2010), researchers gathered data on three timelines that covered constructions built according to changes in household structure and incomes over time. These timelines were later overlapped to examine the interactions among these indicators (Figure 17.3). These graphics were used to identify patterns of incremental construction and to analyse the factors influencing the pace of this process. The software used was the same as in Stage 1.

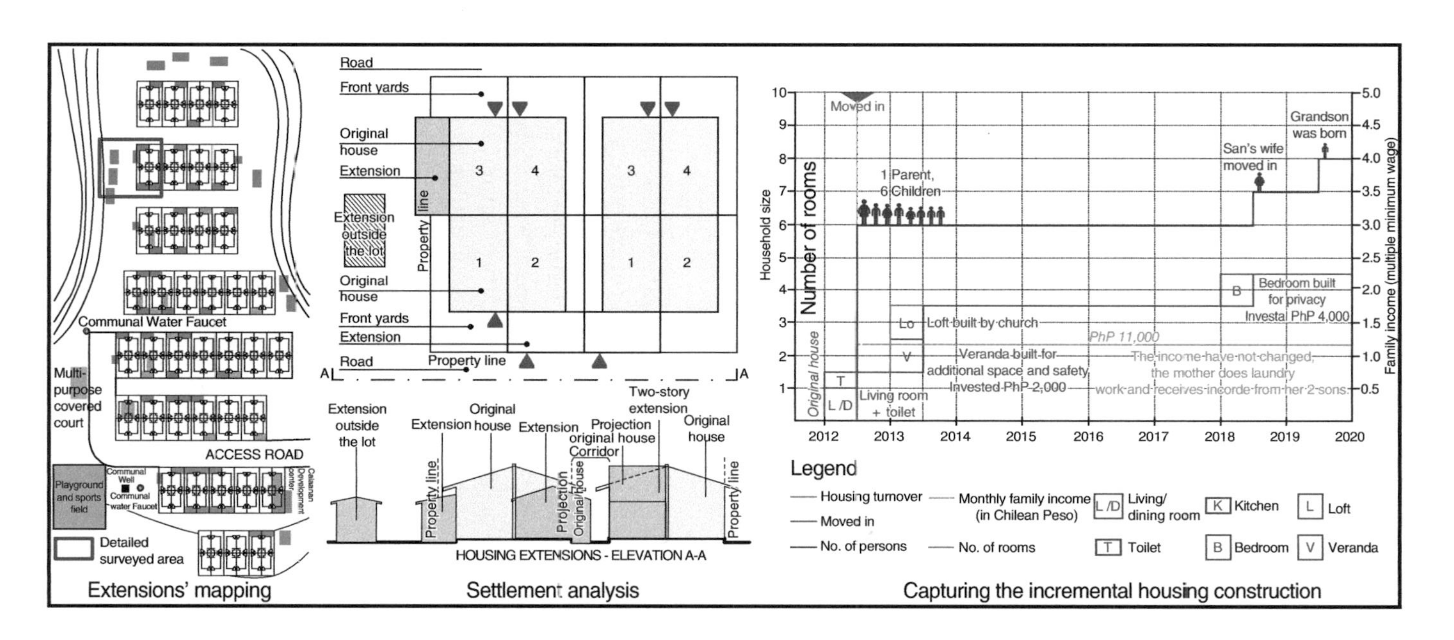

Figure 17.3 Data analysis and outcomes

- Stage 3

  The interviews conducted in 2021 provided evidence of a change in the government's approaches towards resident-built housing extensions being incorporated into the development planning of the resettlement sites. However, government approaches were incipient and did not match the scale and complexities of constructions made by the residents. Integrating the findings in Stages 1 and 2 led to recommendations to address the issues emerging in the incremental housing and settlement development process to minimise the possible risks associated with informal constructions and urban development. The analysis of the data was done using NVIVO software.

Case Study 2: Incentives and Motivations for Enhancing Decision-Making about Earthquake Risk Mitigation in New Zealand

Study Setting

The increasing scale of losses from rapid-onset disasters that do not allow warnings, such as earthquakes, is the result of extensive damage to life, property, and the economy, as well as social disruption. New Zealand has experienced various earthquakes throughout its history but the severity of the 2011 Christchurch earthquake provided evidence of New Zealand's vulnerability to earthquakes. Buildings that have insufficient seismic capacity contribute to the susceptibility of the built environment to earthquake hazards and contribute to losses caused by disaster. Therefore, disasters have reinforced the need for property owners to become proactive in developing programmes to reduce earthquake risk and taking decisions to reduce their level of vulnerability. However, building owners are often reluctant to adopt mitigation measures to reduce losses caused by earthquakes despite advancements in seismic design methods and legislative frameworks. Owners of earthquake-prone buildings (EPBs) are not incorporating appropriate risk mitigation measures in their buildings, as revealed by the large number of buildings that collapsed during the Christchurch earthquake. Owners of EPBs have been found to be unwilling or lack motivation to adopt adequate mitigation measures that will reduce their vulnerability to earthquake risks (Egbelakin *et al.*, 2015).

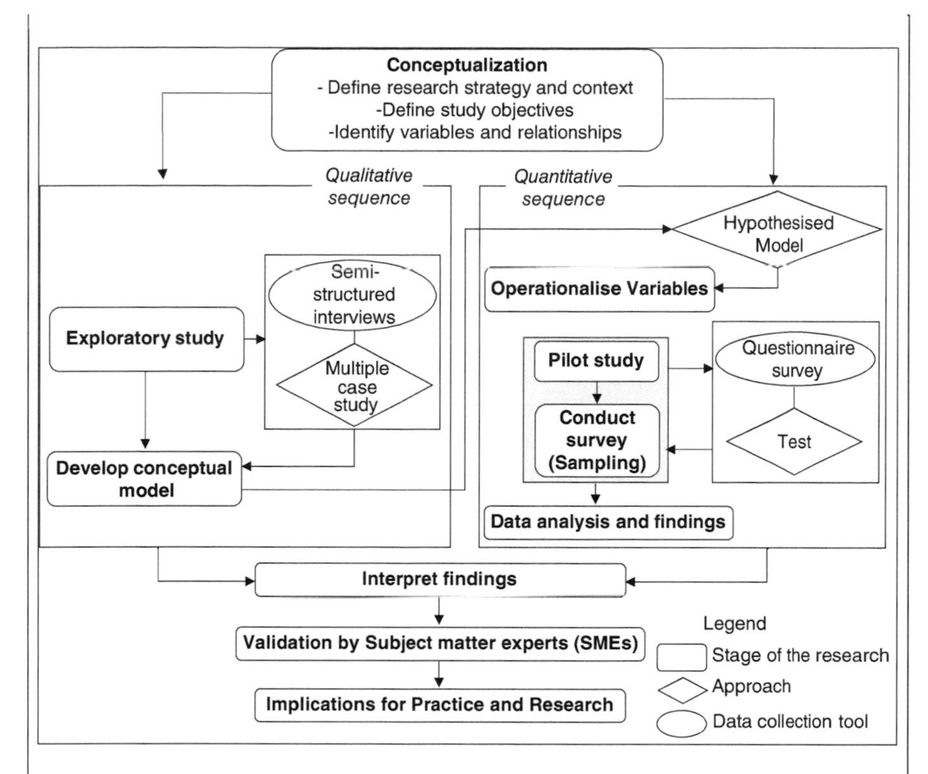

Figure 17.4 Research process case study 2

The investigation in this case study is about how to increase the likelihood of building owners undertaking appropriate mitigation actions to reduce their vulnerability to earthquakes. A sequential, exploratory, two-phase, mixed methods approach was adopted for the research investigation (see Figure 17.4). The approach used in this study was to target four cases in the first qualitative phase (Egbelakin *et al.*, 2013; Egbelakin *et al.*, 2014) and to develop and test a framework in the second quantitative phase (Egbelakin *et al.*, 2017). The different geographic cities selected in the study were Case 1 – Auckland, Case 2 – Christchurch, Case 3 – Gisborne and Case 4 – Wellington. These regions were chosen using a risk-based selection method with criteria such as seismicity, hazard factor and percentage of retrofitted and non-retrofitted EPBs.

Data Collection

The participants of interest in this study were stakeholders involved in making decisions about the mitigation of earthquake risk. The stakeholders identified for this research included building owners, property valuers, engineers and architects, managers of insurance, and financial and governmental organisations involved in earthquake rehabilitation projects. Participants were selected using purposeful sampling based on their knowledge and experience in making decisions about the mitigation of earthquake risk and rehabilitation of EPBs.

- The qualitative phase

  This phase of the research included personal face-to-face interviews because of the relevance of the topic to the stakeholders, and the use of intensive probing questions to gain more insights into the research problem. The objective of the interviews was to investigate the factors affecting decisions made by building owners to adopt earthquake mitigation measures. Furthermore, the interviews informed the analysis of potential intrinsic and extrinsic motivators for decisions made by owners to adopt measures to mitigate earthquake risk. A total of 35 interviews were conducted in the four cities. In addition to semi-structured interviews, some participants provided documents, such as organisation or industry reports on earthquake rehabilitation projects, territorial earthquake policies and insurance, and financial documents of some EPBs, which were helpful during the data analysis stage.

  A semi-structured interview protocol was used to allow for structured spontaneous discussion and follow-up questions on the research topic. The interview protocol was focused on knowledge of earthquake risk, perceptions, and mitigation decisions. The open-ended questions ensured that all research issues were at least considered to enable comparisons among the cases.

- The quantitative phase

  The design of this phase included survey and experimental methods. A survey is the most frequently used mode of observation in studies related to social sciences such as hazard and disaster management (Lindell & Prater, 2000). In addition, experimental methods in social sciences include research conducted in a laboratory environment where human subjects are used. Field and Hole (2002) explained that a researcher has direct control over the research environment and subjects in experimental research through manipulation and randomisation.

  Three sample frames were used to achieve an acceptable level of survey response. The first frame comprised the list of owners and project team members sourced during the qualitative phase. The interview participants were excluded from the survey to avoid duplicate information. The second sample comprised owners of potential EPBs identified after the enactment of the Building Act (2004). The databases of councils developed during the initial evaluation procedure by corresponding territorial authorities were used to extract only the addresses and names of the vulnerable EPBs identified. The councils did not release details of the owners because of privacy issues. Details of the identified owners of potential EPBs were extracted from the certificates of title from the New Zealand properties database. This database was purchased from Property IQ, a commercial organisation that provides online building information to the New Zealand property market. Lastly, the third sample frame comprised members of industry or professional organisations involved in making decisions about the mitigation of earthquake risk and the rehabilitation of EPBs, which was sourced from different industry organisations.

A pilot investigation was conducted with 28 participants to pre-test the efficacy of the questionnaire. An online self-administered questionnaire was for faster geographic coverage, cost-effectiveness, time constraints and to allow the respondents more time to consider the questions before submitting the questionnaire. The survey was open for 12 weeks from March to May 2010. The survey portal made it possible to track incomplete responses and non-responding participants. Respondents who had not replied within three weeks received a follow-up reminder email and a phone call. Another step to achieve an acceptable response was to invite participation through industry and professional organisations.

A total of 510 online questionnaires were administered to the six different stakeholders involved in making earthquake mitigation: building owners, managers of financial institutions, managers of insurance institutions, property valuers and managers, professionals, and government officials. A total of 208 surveys were completed online, representing a response rate of 40.8%.

Data Analysis and Integration

- Qualitative data

Data obtained during interviews was analysed using a cross-case analysis as the focus was on making theoretical generalisations across the cases. A cross-case analysis requires an analysis of the different cases first and, subsequently, the analysis across all cases (Yin, 2013).

The findings from the qualitative phase provided significant insights into the factors affecting decisions about earthquake mitigation made by building owners and possible ways to enhance decisions about the rehabilitation of EPBs. The findings were intended to provide avenues and strategies to reduce earthquake vulnerability in New Zealand. The participants' viewpoints were compared and contrasted collectively, both within each case and across the cases using NVIVO software. The findings from the qualitative data phase were used to develop a framework and generate several hypotheses that were tested in the subsequent quantitative study phase (see Figure 17.4).

- Quantitative data

  Structural Equation Modelling (SEM) was used to analyse quantitative data. In this research, SEM was used to address the research problem by simultaneously estimating the relationships among the theoretical constructs (LVs) and assessing the reliability and validity of the manifest variables (measurement items) of the LVs. SEM enables individual measurements of these constructs based on multi-item scales and hence, was suitable to examine the theoretical framework developed in this study. The data were analysed using SmartPLS2.0 M3 statistical software for SEM analysis. In addition, descriptive statistics and exploratory factor analysis (EFA) were used in this research as supplementary statistical methods. Descriptive statistics were used to describe the properties of central tendency (mean) and variation in the data by computing arithmetic means, variance and standard deviation and, subsequently, to summarise the data. EFA was carried out to examine the proposed dimensionality of the "extrinsic motivator" construct.

  Validation of the overall research findings was undertaken to ensure that the results of the qualitative and quantitative analyses were generalisable. Subject matter experts were used to validate the research findings. Using another sample from the population, researchers conducted face-to-face interviews with five subject matter experts who included two private building owners, two directors from the city councils and one director of a property valuation company.

Discussion

The mixed methods research designs used in the case studies presented in this chapter, as a core methodological approach, incorporated multiple tools for data collection and analysis (see Table 17.1). In this section, the research designs used in both cases are compared following a convergent approach (Creswell & Clark, 2017). The benefits and challenges in each study are interpreted by integrating the findings to examine their convergences and divergences, as summarised in Table 17.1.

Convergences

a Both studies were based on the implementation of mixed methods research applied to multiple case studies from which primary data was collected.

b Multiphase research designs were used in both studies, during which data was collected sequentially, as presented in Table 17.1. In Case Study 2 and the first stage of Case Study 1, qualitative data was collected initially to understand the socio-cultural contexts and the research implications clearly. The results of this initial qualitative phase became the basis for the subsequent quantitative research.

Table 17.1 Comparison of the case studies analysed

	Case study 1	*Case study 2*
Mixed methods design	Mixed methods with multiple case studies Longitudinal study	Mixed methods with multiple case studies
Data collection approaches	Stage 1: QUAL (1 wave) → QUAN (2 waves: exploratory + survey) Stage 2: QUAL (1 wave) + QUAN (1 wave) Stage 3: QUAL (1 wave)	QUAL (1 wave) → QUAN (2 waves: exploratory + model application) → QUAL (1 wave: validation)
Data collection instruments	Semi-structured interviews (face-to-face) Questionnaire survey (face-to-face) Semi-structured interviews (online/video call)	Semi-structured interviews (face-to-face) Questionnaire survey (online self-administered)

QUAL = qualitative; QUAN = quantitative → = sequential + = concurrent () = waves and purpose
Source: Original.

c Following the initial qualitative phases, the hypotheses and frameworks were adjusted for contextualisation prior to the quantitative phases.
d Both case studies included pilot studies as pre-tests in the early stages of the quantitative phases. The results of these pilot studies were the basis for adjustments and operationalisation of the research variables in the quantitative phase.
e Multiple data collection instruments that were adapted based on practicality and efficiency to address research objectives were used in both cases. For instance, researchers in both studies alternated between in-person and online methods for data collection as well as assisted and self-administered surveys.

Divergences

a Both case studies included various phases. However, longitudinal research, developed in three stages over seven years, was used in Case Study 1. The sites investigated throughout the research were revisited and the people's behaviours were re-evaluated throughout the research. The approaches and impacts of the research were also reviewed and adjusted in the subsequent research stages. Conversely, Case Study 2 included quantitative and qualitative phases within a defined timeframe.
b Sequential qualitative and quantitative data collections were used in Case Study 2 throughout the research. However, sequential and simultaneous data collection approaches were combined in Case Study 1, which resulted from the analysis of the findings in previous stages of the longitudinal research (Figure 17.2 and Table 17.1).
c Case Study 2 included the explicit validation of the findings and the analysed framework by engaging relevant real estate professionals and government

officials as subject matter experts. Validation was a core research component in Case Study 2, which provided a basis for generalisation and practical applications.

d Case Study 1 adopted an implicit approach for analysing the impacts and practical implications of the research findings. After analysing the drivers, impacts and challenges for community-driven modifications of the built environment, the research included examining impacts on policy and governance, as seen in Stage 3 of the longitudinal research.

The analysis of the different research stages provided evidence of a process of adjusting the research approaches to enable the fulfilment of the main research objectives of the case studies presented in this chapter. However, as the studies progressed there was space for the adjustment of the operationalisation of the studies and some of their specific research objectives. These variations were necessary to capture the complexities of interactions between people and built environment to build resilience in communities.

Conclusions

This chapter contains an analysis of the different approaches for mixed methods research applied in investigating community resilience and interaction with the adaptation of the built environment. The case studies analysed illustrated the use of mixed methods focusing on adaptability in applying research designs and approaches that is crucial for research about disaster and community resilience throughout the different stages of the research. The analysis of the case studies provided evidence of the importance of integrating different data collection instruments and analysis in which qualitative and quantitative methods were integrated. Furthermore, research in the context of crises, such as disasters, involves challenges of uncertainty and permanent change, access to research participants and on-site data collection. Recently, the need for flexible approaches in mixed research is being revisited in studies using a bricolage methodological approach, triggered by the challenges of the COVID-19 pandemic, as presented in the first section of the chapter.

Therefore, the adaptable nature of research in the context of crises and disasters is the result of the difficulties for researchers in transferring desk-research design to implementation. A thorough understanding of the research problem from the perspectives of stakeholders and participants within a particular context is crucial to the adjusting and improvement of the research design in this discipline. Furthermore, the application of pilot studies provides opportunities for operationalising and further improvement of the instruments used for data collection. Variations in the research process, such as sequential and concurrent are also considered in the research approaches that require revision and updates by researchers. Finally, subsequent variations in data collection instruments and strategies required to address the research objectives were presented in the chapter. However, it is crucial to maintain consistency across the data collection instruments and analysis to maintain high-quality data and rigorous research.

References

Amaratunga, D., Baldry, D., Sarshar, M. & Newton, R. 2002. Quantitative and qualitative research in the built environment: Application of "mixed" research approach. *Work Study*, 51(1):17–31.

Andharia, J. 2020. *Disaster Studies: Exploring Intersectionalities in Disaster Discourse*. Springer Nature, Singapore.

Bosher, L. & Chmutina, K. 2017. *Disaster Risk Reduction for the Built Environment*. John Wiley & Sons, Chichester.

Bueddefeld, J., Murphy, M., Ostrem, J. & Halpenny, E. 2021. Methodological bricolage and COVID-19: An Illustration from innovative, novel, and adaptive environmental behavior change research. *Journal of Mixed Methods Research*, 15(3): 437–461.

Carrasco, S., Ochiai, C. & Okazaki, K. 2016a. Disaster induced resettlement: Multi-stakeholder interactions and decision making following tropical storm Washi in Cagayan de Oro, Philippines. *Procedia-Social and Behavioral Sciences*, 218: 35–49.

Carrasco, S., Ochiai, C. & Okazaki, K. 2016b. Impacts of resident-initiated housing modifications in resettlement sites in Cagayan de Oro, Philippines. *International Journal of Disaster Risk Reduction*, 17: 100–113.

Carrasco, S., Ochiai, C. & Okazaki, K. 2017. Residential satisfaction and housing modifications: A study in disaster-induced resettlement sites in Cagayan de Oro, Philippines. *International Journal of Disaster Resilience in the Built Environment*, 8(02): 175–189.

Creswell, J.W. & Plano Clark, V.L. 2017. *Designing and Conducting Mixed Methods Research*, Sage Publications, Thousand Oaks, CA.

Denzin, N. & Lincoln, Y. 1994. Introduction: Entering the field of qualitative research. In: N.K. Denzin, & Y.S. Lincoln (eds). *Handbook of Qualitative Research*, pp. 1–17. Sage, Thousand Oaks, CA.

Doyle, L., Brady, A.-M. & Byrne, G. 2009. An overview of mixed methods research. *Journal of Research in Nursing*, 14(2): 175–185.

Egbelakin, T., Wilkinson, S. & Ingham, J. 2014. Economic impediments to successful seismic retrofitting decisions. *Structural Survey*, 32(5): 449–466.

Egbelakin, T., Wilkinson, S., Ingham, J., Potangaroa, R. & Sajoudi, M. 2017. Incentives and motivators for improving building resilience to earthquake disaster. *Natural Hazards Review*, 18(4): 04017008.

Egbelakin, T., Wilkinson, S., Potangaroa, R. & Ingham, J. 2013. Improving regulatory frameworks for earthquake risk mitigation. *Building Research & Information*, 41(6): 677–689.

Egbelakin, T., Wilkinson, S., Potangaroa, R. & Rotimi, J. 2015. Stakeholders' practices: A challenge to earthquake risk mitigation decisions. *International Journal of Strategic Property Management*, 19(4): 395–408.

Faber, M.H., Giuliani, L., Revez, A., Jayasena, S., Sparf, J. & Mendez, J.M. 2014. Interdisciplinary approach to disaster resilience education and research. *Procedia Economics and Finance*, 18: 601–609.

Field, A. & Hole, G. 2002. *How to Design and Report Experiments*. Sage, London.

Gilligan, J.M. 2021. Expertise across disciplines: Establishing common ground in interdisciplinary disaster research teams. *Risk Analysis*, 41(7): 1171–1177.

Groat, L.N. & Wang, D. 2013. *Architectural Research Methods*. John Wiley & Sons, Hoboken, NJ.

Johnson, R.B., Onwuegbuzie, A.J. & Turner, L.A. 2007. Toward a definition of mixed methods research. *Journal of Mixed Methods Research*, 1(2): 112–133.

King, D. & Gurtner, Y. 2021. Focusing post-disaster research methodology: Reflecting on 50 years of post-disaster research. *The Australian Journal of Emergency Management*, 36(4): 32–39.

Lindell, M.K. & Prater, C.S. 2000. Household adoption of seismic hazard adjustments: A comparison of residents in two states. *International Journal of Mass Emergencies and Disasters*, 18(2): 317–338.

Magis, K. 2010. Community resilience: An indicator of social sustainability. *Society & Natural Resources*, 23(5): 401–416.

McEachan, R., Dickerson, J., Bridges, S., Bryant, M., Cartwright, C., Islam, S., Lockyer, B., Rahman, A., Sheard, L., West, J., Lawlor, D., Sheldon, T., Wright, J., Pickett, K. & Null, N. 2020. The born in Bradford COVID-19 research study: Protocol for an adaptive mixed methods research study to gather actionable intelligence on the impact of COVID-19 on health inequalities amongst families living in Bradford. *Wellcome Open Research*, 5: 191.

McSweeney, M. & Faust, K. 2019. How do you know if you don't try? Non-traditional research methodologies, novice researchers, and leisure studies. *Leisure/Loisir*, 43(3): 339–364.

Riazi, A.M. & Candlin, C.N. 2014. Mixed-methods research in language teaching and learning: Opportunities, issues and challenges. *Language Teaching*, 47(2): 135–173.

Rodríguez, H., Quarantelli, E.L., Dynes, R.R., Andersson, W.A., Kennedy, P.A. & Ressler, E. 2007. *Handbook of Disaster Research*. Springer, New York.

Samah, A.A., Zaremohzzabieh, Z., Shaffril, H.A.M., D'Silva, J.L. & Kamarudin, S. 2019. Researching natural disaster preparedness through health behavioral change models. *American Journal of Disaster Medicine*, 14(1): 51–63.

Sharp, H. 2019. Bricolage research in history education as a scholarly mixed-methods design. *History Education Research Journal*, 16(1): 50–62.

Suárez, M., Gómez-Baggethun, E. & Onaindia, M. 2020. Assessing socio-ecological resilience in cities. In: M.A. Burayidi, A. Allen, J. Twigg & C. Wamsler, (eds). *The Routledge Handbook of Urban Resilience*, pp. 197–216. Routledge, London.

Tashakkori, A. & Teddlie, C. 2010. *SAGE Handbook of Mixed Methods in Social & Behavioral Research*. 2nd edition. SAGE, Thousand Oaks, CA.

Tavares, I.M., Fernandes, J., Moura, C.V., Nobre, P.J. & Carrito, M.L. 2021. Adapting to uncertainty: A mixed-method study on the effects of the COVID-19 Pandemic on expectant and postpartum women and men. *Frontiers in Psychology*, 12: 1–13.

Taylor, R., Forrester, J., Pedoth, L. & Matin, N. 2014. Methods for integrative research on community resilience to multiple hazards, with examples from Italy and England. *Procedia Economics and Finance*, 18: 55–262.

Tight, M. 2017. *Understanding Case Study Research: Small-scale Research with Meaning*. Sage, London.

Timans, R., Wouters, P. & Heilbron, J. 2019. Mixed methods research: What it is and what it could be. *Theory and Society*, 48(2): 193–216.

Twigg, J. 2015. *Disaster Risk Reduction*. Humanitarian Practice Network, London.

Witt, E. & Lill, I. 2018. Methodologies of contemporary disaster resilience research. *Procedia Engineering*, 212: 970–977.

Yin, R.K. 2013. *Case Study Research: Design and Methods*. Sage Publications, Los Angeles, CA.

Zeisel, J. 1984. *Inquiry by Design: Tools for Environment-Behaviour Research*. Cambridge University Press, Cambridge.

Index

Note: **Bold** page numbers refer to tables and *italic* page numbers refer to figures.